Human Freedom in the Age of AI

This book claims that artificial intelligence (AI) may affect our freedom at work, in our daily life, and in the political sphere. The author provides a philosophical framework to help make sense of and govern the ethical and political impact of AI in these domains.

AI presents great opportunities and risks, raising the question of how to reap its potential benefits without endangering basic human and societal values. The author identifies three major risks for human freedom. First, AI offers employers new forms of control of the workforce, opening the door to new forms of domination and exploitation. Second, it may reduce our capacity to remain in control of and responsible for our decisions and actions, thereby affecting our free will and moral responsibility. Third, it may increase the power of governments and tech companies to steer the political debate, thereby affecting the possibility of a free and inclusive political participation. The author claims that it is still possible to promote human freedom in our interactions with AI. This requires designing AI systems that help promote workers' freedom, strengthen human control and responsibility, and foster a free, active, and inclusive democratic participation.

Human Freedom in the Age of AI will be of interest to scholars and graduate students working on the ethics of technology, philosophy of technology, political philosophy, design, and artificial intelligence.

Filippo Santoni de Sio is Associate Professor in Ethics and Philosophy of Technology at Delft University of Technology (The Netherlands) and Associate Editor at the *Journal of Ethics*. He has co-authored more than 60 academic papers, two monographs, and three edited books, including the recent *Research Handbook on Meaningful Human Control of Artificial Intelligence Systems* (2024).

Routledge Research in Applied Ethics

Why Conscience Matters
A Defence of Conscientious Objection in Healthcare
Xavier Symons

The Ethics of Agribusiness
Justice and Global Food in Focus
Shane Epting

What Kind of Death
The Ethics of Determining One's Own Death
Govert den Hartogh

Human Enhancement and Well-Being
A Case for Optimism
Emma C. Gordon

Urban Enlightenment
Multistakeholder Engagement and the City
Shane Epting

Moral Challenges in a Pandemic Age
Edited by Evandro Barbosa

The Ethics of Drone Design
How Value-Sensitive Design Can Create Better Technologies
Dylan Cawthorne

New Omnivorism and Strict Veganism
Critical Perspectives
Edited by Cheryl Abbate and Christopher Bobier

Human Freedom in the Age of AI
Filippo Santoni de Sio

For more information about this series, please visit: https://www.routledge.com/Routledge-Research-in-Applied-Ethics/book-series/RRAES

Human Freedom in the Age of AI

Filippo Santoni de Sio

Routledge
Taylor & Francis Group
NEW YORK AND LONDON

First published 2024

by Routledge
605 Third Avenue, New York, NY 10158

and by Routledge
4 Park Square, Milton Park, Abingdon, Oxon, OX14 4RN

Routledge is an imprint of the Taylor & Francis Group, an informa business

ISBN: 978-1-032-30054-2 (hbk)
ISBN: 978-1-032-30056-6 (pbk)
ISBN: 978-1-003-30324-4 (ebk)

DOI: 10.4324/9781003303244

Typeset in Sabon
by Deanta Global Publishing Services, Chennai, India

Contents

Preface

Recall the Mesopotamian parable of the appointment in Samarra. A servant goes to the market in Baghdad, where he spots a mysterious lady in the crowd insistently looking at him. As he immediately recognises Death in her, he asks his master to lend him a horse, jumps on it, and rides away as fast and for as long as he can. Exhausted, he eventually stops in Samarkand to find the lady waiting for him: 'thankfully you made it to our appointment, you were so far away a couple of days ago.'

Sometimes it is difficult to escape one's destiny. My mother works as a lawyer in my hometown Torino (Italy) and my father did the same until he passed away two years ago. When I was approaching the end of school, I thought that my destiny would be to follow in their professional footsteps, even though I was not fully convinced of that choice. My friends and family encouraged and helped me in devising my first escape from my destiny: following my passion and taking a degree in philosophy. Destiny knocked back at the door though when it was time to find a topic for my PhD project: ironically, I ended up writing a dissertation on the philosophical foundation of legal responsibility.

Joking aside, while I wholeheartedly endorsed and never regretted the choice of my PhD topic, the unwanted destiny I started to feel hanging over me as a wannabe academic philosopher was that of a boring life of abstract studies, isolated from reality and people, and unable to have any meaningful interactions with them. Thankfully, another escape was waiting for me around the corner. Due to the lack of concrete opportunities for an academic career in Italy, I got my diploma as a high-school teacher in philosophy and history and taught at a high school for two and half years. Teaching and education were defining experiences in my realisation of what philosophy and humanities can and should be – a powerful means to engage people in conversations on things that matter; a constant commitment to sharpen concepts and theories to capture reality and, whenever possible, to change it. Also, the fun, warmth, and challenges of human relationships. I have kept trying to bring this attitude to the classroom, also in my successive experiences of teaching ethics and philosophy to engineering students in Delft.

Looking back at those first years of my career, I want to acknowledge and thank my family and friends for helping me find my way in life and in academic studies; I am deeply grateful to them, to my students at school and at the university, and above all to my wife and life partner Elena and to our son Leone, for reminding me every day that the most important, beautiful and challenging things in life happen in the world outside philosophy books. And it is this world that, also as an academic philosopher, I want to engage, learn from, and talk to.

I encountered the biggest opportunity to give my professional destiny a twist in 2011, when Nicole A Vincent invited me to apply for a position as a postdoc in ethics of technology at Delft University of Technology in The Netherlands, in the relatively new but fast-growing and ambitious Department of Ethics and Philosophy of Technology, where she was working as a researcher. I joined TU Delft in 2012. I am grateful to Nicole for inviting me to make the move that would define my academic career for the years to come, for introducing me to the ethics of emerging technologies, and for generously connecting me to her international network of scholars.

Among Delft's scholars was Jeroen van den Hoven. Over the past 12 years, Jeroen has been the person who more than anyone helped me find my way in the world of applied ethics of technology, opening doors to collaborations with top researchers and high-level policymakers, suggesting topics and directions for my research and involving me in plenty of events and initiatives. His competence and enthusiasm have been a constant drive for my research. His commitment to what he proudly calls the 'Delft approach' to ethics and philosophy – one that is responsive to technical and societal problems and that utilises philosophy to clarify and address them 'by design' – has been a major inspiration for my work in general and for this book in particular. This is where the 'design stance' advocated for in this book comes from.

To Jeroen I also owe the idea of 'Meaningful Human Control' over AI, which he first invited me to co-develop for his statement at the 2015 United Nations meeting on the Convention on Certain Conventional Weapons (CCW) in Geneva. We later developed those ideas in a 2018 co-authored article titled 'Meaningful Human Control over Autonomous Systems: A Philosophical Account,' which Chapter 8 of this book partly reproduces. Jeroen also encouraged me to address the issue of AI at work, which we first explored in a 2021 paper co-authored also with Txai Almeida: 'The Future of Work: Freedom, Justice and Capital in the Age of Artificial Intelligence.' The paper was inspired by previous conversations with another Delft colleague, the economist Ro Naastepad. I owe some of the ideas on AI and work presented in this book – the neo-republican analysis of Universal Basic Income in Chapter 1, the exploitation of mental activities by tech companies presented in Chapter 5, and the analysis of 'spheres of justice' of Chapter 6 – to that paper and its preparation. Additional insights on the ethics of robots at work came from my collaboration with Madelaine Ley in the Delft AiR Lab project, as well as David Abbink and other researchers in the Worker-Robot Relations consortium (WRR).

I further developed the concepts of Meaningful Human Control and 'responsibility gaps' with Giulio Mecacci and the other members of the NWO-funded multidisciplinary project 'Meaningful Human Control over Automated Driving System' at TU Delft, as well as with members of the Delft AiTech program. These projects have been another great source of inspiration for my attempts at connecting philosophy and engineering. With Giulio Mecacci I co-authored the 2021 article: 'Four Responsibility Gaps with Artificial Intelligence: Why they Matter and How to Address Them,' which Chapter 7 of this book partly reproduces.

My interest in and knowledge of the neo-republican approach to freedom, and the potential for its application to AI systems, partly derives from my collaboration with Marianna Capasso on her PhD thesis *Positions of Power. Freedom and Control in the Age of Artificial Intelligence*. My interest in Critical, Speculative, and more-than-human Design was partly sparked by some conversations with the Delft colleagues and design scholars Maria Luce Lupetti and Dave Murray-Rust. Literature research on some of the book topics was done as part of their internships at TU Delft by Daniëlle Smit, Kelly Mink, Lotte van Elteren, and Luna van Dijk. Many other ideas contained in the book come from conversations and encounters that I cannot recall. Maybe someone will recognise them while reading the book – thank you to you too.

In the past years I had the opportunity and the honour to collaborate in policy initiatives at the level of Dutch government (White Paper on Ethics and Self-Driving Cars), the European Commission (H2020 Expert Group on Ethics of Connected and Automated Vehicles), and the United Nations (meetings of the UNECE WP1 on road traffic safety). These experiences of close interaction with people from other disciplines with stronger competencies in technical and policy matters have also greatly helped me broaden and strengthen my approach to the ethics of AI.

When I started working on this book more than two years ago, I was not fully aware of how challenging this project would be. At various stages of its preparation, I greatly benefitted from the patient advice and critical feedback of many colleagues and friends who kindly accepted to read my drafts, listened to my ideas, and provided written comments along the entire preparation process, from the editorial proposal until the final drafts. These exchanges have been among the most rewarding and precious moments of personal growth in this process. This book would have been simply impossible without you folks! Thank you to Aarón Moreno

Inglés, Aishwarya Suresh Iyer, Antonia Sattlegger, Elena Bisagni, Giuliano Bracci, Giulio Mecacci, Ibo van de Poel, Kritika Maheshwari, Janna van Grunsven, Jelena Belic, John Danaher, Jonne Maas, Joost Mollen, Jordi Viader Guerrero, Luciano Cavalcante Siebert, Mahan Mehrvarz, Marianna Capasso, Marie-Therese Skewenz, Matthew Dennis, Nelly Raftopoulou, Sabine Roeser, Sofie Dideriksen, Steffen Steinert, Stephen Rainey, Tim Christiaens, and Udo Pesch.

Thank you also to two anonymous reviewers who assessed and commented on the book proposal for Routledge, and all the many others whom I have forgotten to mention. Thank you to Diederik Rep for the incredible editorial work on the final draft; and to the Delft Digital Ethics Centre and the Section Ethics and Philosophy of Technology at Delft University of Technology for generously funding the professional editing of this book.

The Hague, 17 October 2023

1 Introduction

Artificial intelligence and the future of freedom

1.1 Human freedom and artificial intelligence

Artificial intelligence (henceforth AI) is a term covering technologies that aim to replicate human intelligence and behaviour and, more generally, solve problems with relatively low direct control by human controllers. ChatGPT, the popular application of AI for text production, is one famous recent example. However, AI has been around for a long time and AI systems can increasingly be deployed in the control of vehicles, medical devices, factories, defence, financial systems, and more.

The fast development of AI in the past years has raised questions and concerns. Can AI and robotics applications take our jobs, leaving us unemployed? Can data-driven automated systems in military or finance become so complex and autonomous that they will escape human control and responsibility? Can privately owned social media platforms and search engines increasingly create information bubbles and spread false information, making it more difficult for people to participate in politics?

Or, on a more optimistic side, can AI and robotic applications free us from dull, dirty, or dangerous jobs, opening better opportunities to actually enjoy life? Can complex systems make our critical societal infrastructures – transportation, defence, and healthcare – safer and more efficient? Can digital technologies eventually create a more open and inclusive democratic participation and even support resistance to authoritarian governments?

In this book I propose to address these and similar questions through the lens of the philosophical and political concept of human freedom. I will try to answer two questions:

> *How can we best conceptualise the impact of AI systems on human freedom?*
>
> and
>
> *How can we design AI system that promote rather than harm human freedom?*

DOI: 10.4324/9781003303244-1

I will address the first question by analysing some key examples of AI systems through the lens of *philosophical theories* of freedom, and I will address the second question by looking at *design theories* with an explicit ethical and political stance towards AI and technology in general.

This introduction briefly illustrates the philosophical conception of freedom grounding the book; how this connects with the themes of justice, responsibility, and democracy, and how a 'design stance' in ethics of technology may contribute to promoting freedom. This introduction also offers an overview of the topics covered in the book.

1.2 Freedom as non-domination

Freedom is one of the most controversial topics in philosophy. While everybody tends to agree on the value of freedom, we tend to disagree on what freedom means. Western modern philosophy alone contains countless accounts of freedom, often in contrast with each other: from rationalist philosophers like Immanuel Kant to socialist revolutionaries like Karl Marx to free-market economists like Frederik Hayek, many modern philosophers have engaged in reflections and pleas on the value of human freedom.

In a seminal article first published in 1958, the philosopher Isaiah Berlin elaborated on a distinction originally proposed by Kant to define *Two Concepts of Liberty*.[1] According to Berlin, two main concepts of freedom – or as he calls it: liberty – are in tension with each other in Western political philosophy. 'Negative liberty' is the possibility to act without interference from other people, other groups, or the State. 'Positive liberty' is the realisation of one's true or deeper self, achieved by actively taking control of one's life and realising one's fundamental purposes. It is connected to the idea of moral autonomy.[2] Positive liberty requires taking a critical distance from one's desires and wishes and identifying those that one really wants or needs to pursue. For instance, I may tend to behave in ways that upon reflection do not match what I consider to be my deeper or better self, or the person that, all things considered, I would like to be or I think I should be. To be morally autonomous – that is, pursuing positive freedom – I should free myself from undesirable habits, wishes, or behaviours, and embrace my better self. This is indeed the concept of freedom embraced by many theologians (freedom from evil and sin), moral philosophers like Kant (rational autonomy), and radical political philosophers like Rousseau and Marx (freedom from unjust social structures).

While positive liberty may be a valuable moral ideal, Berlin was concerned about the risks of its translation into a political ideal. What if we start to say not only that it is good that people develop their moral autonomy or positive liberty but that they *should* do so. This could lead to the identification of a universal criterion or even the creation of an external authority that can tell people which version of themselves they should favour to achieve moral autonomy. This is the risk of moralism or 'perfectionism.' A possible next step is forcing people, through the intervention of the State, to pursue this better or deeper self. This is the risk of 'paternalism' or even authoritarianism: political oppression realised in the name of moral freedom.[3] Berlin was writing in the time of the Cold War and had in mind the Soviet Union, where many people were enslaved in the name of their liberation from the alienation of capitalism. But is the risk of political oppression sufficient to stick to a minimal conception of political freedom as non-interference, giving up on the ideal of positive freedom or moral autonomy?

Recently, Philip Pettit and others have proposed a third way between negative and positive freedom: 'freedom as non-domination.'[4] According to Pettit, not only action-interference but also mere imbalances of social powers make people unfree; not only because they make it more likely that those in power will interfere but also because they make people vulnerable and subject to the arbitrary will of the others, thereby affecting their *status* of free agents.[5] This approach, which Pettit calls 'republican,' is more often referred to as 'neo-republican.'

Neo-republicanism takes a normative stance towards social structures and relationships, in addition to interactions between people. It integrates the concept of political freedom with the idea of social justice or at least social equality. There cannot be freedom in a society where some agents can exercise an uncontrollable or arbitrary power over other agents. To be clear, power structures should not be eliminated to prevent domination (this would probably be impossible), but all powers must be controllable, meaning that all those who are subject to other people's power must have the possibility to contest it. The analysis of freedom requires an analysis of the power structures in society: which social structures allow for domination, and which protect from domination and promote freedom?

One consequence is that freedom as non-domination cannot be realised outside democracy, as, by definition, in the absence of democratic institutions, governmental powers cannot be controlled and people would therefore be subject to domination. Interestingly, negative freedom can instead be realised, in principle, also in a dictatorship, whenever the government decides not to interfere with someone's choices and actions.[6]

Another important consequence is that political freedom does not only depend on the form of government but also depends on other power structures in society. It is possible for someone to be *un*free in a democratic country if, for instance, they are subject to the uncontrollable will of an employer, women to the arbitrary will of men, some ethnic groups to the uncontrollable will of others, people without financial resources to the will of the wealthier. Freedom as non-domination is an ideal of political freedom broadly conceived. It is, as it were, an ideal of moral or social freedom.

Freedom as non-domination strikes an interesting balance between negative and positive freedom. It requires more than negative freedom, because in addition to preventing actual interference in the choices and actions of people, it also forbids potential interference, that is, imbalances of power that allow groups or people to exert their uncontrollable will on others, making them vulnerable and dependent on the more powerful. At the same time, it does not bring the risks of oppression of positive freedom, because, by definition, the theory does not allow anybody to deprive others of freedom. Also, it does not even bring the risk of moralism or perfectionism insofar as it does not require people to pursue moral autonomy or positive freedom – even though it does prescribe that everybody remains free to do so, if they want and can.[7]

The neo-republican concept of freedom as non-domination is an important starting point for this book.[8]

A first claim I will make is that AI is a set of technologies produced by and operating as a part of what I will call a socio-technical system, a set of social, political, economic, and cultural norms and practices. As I will argue, AI is not only a technology but also a part of a power structure.

A second claim is that, as a socio-technical system, AI can increase the domination of some agents or groups on others, in work relationships as well as in other social and political relationships.

A third claim is that as a socio-technical system combining technical and social elements, AI and its impact are potentially open to be shaped and steered in different directions that can protect or promote freedom. This will however require taking what I will call a 'design stance' aiming to systematically change the ethical, social, political, and cultural principles driving its development, design, and use. The next sections briefly elaborate on these three points and present in more detail the content of the book.

1.3 The politics of technology

In an essay from 1980, the sociologist and philosopher of technology Langdon Winner famously stated that 'artifacts have politics.'[9] Artefacts (meaning objects and technological systems) do not happen in a void. They do not appear, as it were, in our hands as neutral tools that we can freely and indefinitely shape and use as we wish. They are always produced and developed as part of social, economic, and political systems whose values and principles they tend to reflect and perpetuate. In one iconic example by Winner, the New York urban planner Robert Moses designed an overbridge with the intention to block a public transport connection and therefore prevent mostly Black people from reaching Long Island. A racist attitude was designed into the bridge and then supported and reinforced by it.[10] As Winston Churchill once reportedly said, in a plea for the reconstruction of the House of Lords bombed out in 1941: 'We shape our buildings; thereafter they shape us.' Technology is shaped by – and contributes to shape – moral and political structures.

In its narrow sense, AI refers to the algorithms and computational methods supporting its problem-solving capacities. In a broader sense, AI is a system that includes the data needed to train the algorithms, the hardware to store the data and to run the software, the interactions with the people using the systems, and more. AI is also a production system with specific material requirements. In her *Atlas of AI*, Kate Crawford presents and documents the massive extraction of data, labour, and natural resources on a global scale to support the current production of AI in industrialised countries.[11]

In this book I will always refer to 'AI systems,' as opposed to just 'AI,' to emphasise the complex network of material, cultural, economic, political, and social-psychological factors inevitably embedded in the development, production, and use of the technology. I will also often refer to AI as being part of a 'socio-technical system.'

The culture or 'narratives' accompanying AI are also arguably part of its socio-technical system, and they may impact human freedom as well. For instance, AI systems are often presented as 'autonomous' because of their capacity to operate with relatively low direct control from human controllers. However, 'autonomous technology' is a dangerous misnomer: misnomer because it obfuscates the extensive network of human research, labour, interactions, language, data, and maintenance that make and sustain its functioning;[12] dangerous because by making the human element invisible, it provides the false image of 'a future that doesn't need us,'[13] thereby legitimising the exploitation of people, making control and responsibility for the development and deployment of AI systems more difficult to achieve, and disincentivising the active political participation of people in shaping the future of technology.

Do we live in 'the age of AI'? We probably do, but not in the sense some think, who expect that 'autonomous AI' will solve (or destroy) everything. It is rather an era filled with *stories* about AI solving everything or making everything unsolvable. Avoiding naïve techno-optimism of technology rescuing the world on its own and dystopian imaginaries of robots taking over, I propose a shift of focus from machine autonomy to human freedom: what does it mean to remain or become free, responsible, and politically engaged human agents in 'the age of AI,' and what does it take to design future systems that will promote human freedom, responsibility, and democracy?

1.4 No freedom without justice, responsibility, democracy, and design

Human freedom is always intertwined with and dependent on the social and socio-technical systems in which people find themselves to act and interact. Therefore, the concept of freedom must be analysed in connections with other three ideas: *justice*, which is not being dominated by other people or groups through AI systems; *control and responsibility,* which is maintaining some meaningful capacity and opportunity to steer technical systems and thereby remaining responsible for their behaviour; *democratic participation*, which is being able to contribute to the broader political life of one's community and to control governments and other powers; *design*, which is the capacity to imagine and realise socio-technical systems that promote freedom.

The topics presented in the book can be clustered around six main themes.

(1) *Freedom requires imagining and designing ethically better socio-technological systems.*

Karl Marx once famously wrote: 'The philosophers have only interpreted the world in various ways; the point is to change it.'[14] This aphorism nicely captures one important, overarching theme of this book. Promoting freedom in the age of AI requires as a first step resisting 'technological determinism,' which is the belief that technological development is an independent variable that cannot be governed and to which people can only adapt or react. Such an attitude is implicit in the many calls to try and predict the direction in which AI and technology will go, to be able to prepare for that. It is also sometimes implicitly accepted in well-intended political proposals like that of a Universal Basic Income to counterbalance the future impact of technology on employment opportunities. However, as future studies scholar Jim Dator once wrote: 'The future cannot be predicted, because the future does not exist.'[15] There are many possible futures, and which one will materialise also depends on what our visions and actions will be. A first exercise of freedom is therefore resisting a view of technology as necessarily driven by the current dominant values of global tech capitalism: indefinite profit, efficiency and massive extraction of data, labour and material resources, and replacement of human workforce (Chapter 2).

Taking a pro-active *design stance* may make alternative visions emerge and realise. The movements of Responsible Innovation[16] and Value-Sensitive Design[17] have proposed and started realising models of technological development and design, where so-called 'stakeholders' – i.e. actors who will be affected by technological development – are systematically included in the technological process, and diverse ethical and political values are proactively designed into technological systems through multidisciplinary projects involving engineers, social scientists, philosophers, lawyers, and others. Another family of design approaches striving to change the dominant approach to technology is that of Critical, Speculative, Social, or Participatory Design. The critical and speculative element consists in using objects and systems to prompt a social and critical reflection about the (un) desirability of present technological approaches and imagining alternative futures.[18] The social and participatory dimension consists of going beyond mere 'stakeholder involvement' by researchers and designers to ensure the technical and institutional empowerment of communities in developing and realising their own projects[19] (Chapter 3).

(2) *Domination of workers* and *exploitation of internet users* must be made visible and addressed.

Freedom crucially depends on the nature of social structures of power, and the extent to which people subject to others' power possess means to control and contest it, thereby avoiding domination. AI systems may support new forms of domination or reinforce existing ones. Workers in highly automated physical spaces like the Amazon warehouses in the US, or workers in the so-called gig economy, who depend on Uber's or other tech company's apps for their job, may not have options to control or contest the policies and decisions of their employers. Decisions about the direction, assessment, reward, or replacement of the workforce are mediated by technical systems whose functioning is largely controlled by the companies and that remain largely opaque and inaccessible to workers.[20] The same may happen in many other, more traditional workplaces.[21] Structural strategies to design for workers' freedom must be devised and supported, and a structural change in the research on the future of work must be pursued (Chapter 4).

Broadening the scope of the analysis beyond traditional employer–employee relationship, owners of internet platforms are also in a position of domination over their services' users because users are increasingly dependent on their platforms for many vital aspects of their lives: work, social relations, information, transportation, financial services, and more.[22] And they cannot sufficiently control or contest the companies' decisions about the functioning of the platforms. This dominating position allows tech companies to exploit and commodify the knowledge and social connections produced on their platforms[23] and the cognitive, social, and emotional work of the platform users.[24] These are some key features of what recent Marxist and neo-Marxist literature has called 'cognitive capitalism.'[25] The exploitation of the users' data is often summarised in the saying: if you are not paying for the product, then you are the product. The exploitation of the users is summarised in the saying: if you are not paying for the product, then you are the worker. Social and political strategies and policies to resist these domination and exploitation must be devised: rather than just focusing on Universal Basic Income, we must also strive for a broader 'universal basic freedom' (Chapter 5).

(3) *Freedom requires value pluralism.*

The protection of people from domination at work and beyond must be combined with the respect of people's different legitimate ethical approaches to work and life. The introduction of AI and robotics in the workplace should avoid the risks of 'paternalism' and 'perfectionism.' In other words, it should not rely on universalistic ethical assumptions about the 'goods of work,'[26] like, for instance, the idea that repetitive work is necessarily bad and its automation necessarily good (Chapter 6).

Pluralism also requires that the different general shared societal goods attached to various professions are not harmed but ideally promoted by the introduction of AI systems. For instance, the introduction of AI and robotic systems in warfare should not affect the possibility of military operations to remain subject to public scrutiny. In Michael Walzer's spirit, pluralism requires that AI respects the separation between 'spheres of justice.'[27] In the example of warfare, the logic of technological (and military) efficiency should not excessively contaminate the logic of public accountability of the defence operations. The same holds *mutatis mutandis* for other professional domains like healthcare or scientific research (Chapter 6).

(4) Freedom requires *Meaningful Human Control* over AI systems and *human responsibility* for their behaviour.

Users and other relevant agents – designers, controllers, and regulators – should be put in a position where they can remain ultimately morally responsible for the system behaviour, no matter how much operational control is delegated to the technical systems.[28] Responsibility gaps should be avoided.[29] Responsibility gaps are situations in which AI systems cause unwanted harm for which there is no excuse or justification (like, for instance, the killing of a pedestrian by an Uber self-driving test car in Arizona five years ago), while at the same time none of the relevant human agents involved can legitimately be held fully accountable or responsible for it, due to lack of sufficient knowledge, control, moral awareness of the risks of the system, or commitment to prevent them (Chapter 7).

Responsibility gaps may be avoided by designing for what I call 'Meaningful Human Control', according to two requirements: the system should be aligned with the values, reasons, and intentions of the relevant human agents ('tracking') and with the capacities of the relevant human agents ('tracing') (Chapter 8).

To grant Meaningful Human Control to all, we must explicitly consider the issues of social justice, in particular, 'epistemic injustice' in the development and use of technology.[30] Epistemic injustice happens when power structures limit oppressed people's opportunity to make their reasons matter in social interactions, including those guiding technological development, and to develop or exercise some relevant cognitive and social capacities, including those required to keep control of technological systems (Chapter 9).

(5) Freedom requires *design for democracy.*

Technology should be used to promote rather than limit *democratic participation.* Much scientific and philosophical literature has recently discussed the (negative) impact of AI-driven content personalisation and spreading of (false) information on privately owned internet platforms on democratic debates and deliberation. A less-explored topic is what researchers, designers, and policymakers can do to promote constructive and inclusive democratic participation instead. On the one hand, following the ideal of 'deliberative democracy' and in line with the ideal of Value-Sensitive Design, policymakers and designers have created the so-called 'e-government' or 'open government' platforms.[31] Platforms may be designed to create a digital space that, unlike existing mainstream privately owned social media platforms, favour rational and constructive engagement grounded in accurate information and even collaborative legislative drafting. On the other hand, following the pragmatist approach of democracy as 'experimentation and inquiry,'[32] designers have applied the ideal of Critical, Social, and Participatory Design to support and coordinate existing bottom-up projects aimed at engaging specific communities in experimenting new forms of democratic participation and action; for instance, a community-based project for the maintenance of the built environment in low-income neighbourhoods.[33] These two trends must be encouraged and expanded (Chapter 10).

(6) Freedom must be *freedom for all.*

Striving towards a more inclusive political freedom requires expanding the idea of democratic participation in four directions. First, expanding the idea of the 'public sphere' to include more systematically non-argumentative forms of contributions,[34] namely aesthetic or expressive forms of engagement, and supporting projects that may contribute to democratic debates via aesthetic engagement, e.g. new media art ('more-than-argumentative democracy'). Second, expanding democratic participation from deliberation to 'anti-corruption'[35] and 'anti-power'[36] and designing systems that may support counter-powers also in non-democratic, authoritarian political regimes ('more-than-representative democracy') (Chapter 11).

Third, inclusivity requires reinforcing the reflection on the colonial aspects of existing digital technologies and supporting decolonial projects and design practices[37] ('post-colonial democracy'). Fourth, it requires extending our understanding of democratic representation to cover the interests and possibly the rights of the natural environment[38] ('more-than-human democracy') and moving from 'human-centred' to 'more-than-human design'[39] (Chapter 12).

1.5 The future of freedom: ethics, politics, and design

While I have the ambition to cover many aspects of human freedom in relation to AI and digital technologies, I do not aim to cover them all. To name just a few, I do not address the issues of digital technology and manipulation,[40] privacy,[41] surveillance,[42] and well-being.[43]

One call this book aims to answer is that of shifting focus from the ethics to the politics of AI. This call has taken two forms: one, more radical, proposes to *replace* the ethical reflection on AI with political reflection and is summarised in the motto: 'Don't ask if artificial intelligence is good or fair, ask how it shifts power.'[44] This phrase expresses the concern that the ethics of AI approach, as exemplified in initiatives such as 'AI for good'[45] or 'fair, accountable, transparent AI,'[46] misses the point, as it does not sufficiently address the broader social and political factors that may make technology development and use unjust or otherwise politically problematic. In this view, discussion over ethical principles might even become harmful when used by companies or political actors to increase their reputation while escaping harder regulations. This is the phenomenon of so-called 'ethics washing.'[47]

While I am very sympathetic to these concerns, I believe the way to address them is not by replacing AI ethics with AI politics. It is instead to *expand* or *integrate* AI ethics with more political philosophy. This is an ongoing trend.[48] However, while the issue of 'algorithmic injustice' has already received widespread attention,[49] this book covers the less-explored topic of AI and human freedom.

Also, the book proposes to take a 'design stance' with regard to the problem of freedom, that is, to take seriously the idea that human freedom can be protected and promoted through a more widespread deployment of design methods such as Value-Sensitive Design and Critical, Speculative, Participatory, and Social Design. Once again, I am not proposing to replace other forms of political action or regulatory interventions, for instance, a stronger control on the political and economic power of big tech or a stricter legal regulation on the use of AI in security or public administration. The two lines of interventions are not incompatible and must be pursued in tandem. For instance, while I believe that stronger redistributive measures such as Universal Basic Income are insufficient to address the impact of AI on the freedom of people, I remain sympathetic towards them. Interestingly, methods like Critical and Speculative Design explicitly recognise the continuity and complementarity between design practices and other forms of political actions. As designer Carl DiSalvo puts it, 'often times design experiments [...] identify and articulate problems that should be addressed, perhaps by people other than designers.'[50]

While current political and economic frameworks do not seem to leave much space for realising alternative visions, we cannot expect a change to happen any time soon if we are not even clear about how these alternatives may look like. Imagining and defining a desirable future direction for technology and design is a necessary – if certainly not sufficient – condition to make that future happen. If we as philosophers, designers, and citizens stop believing in the power of human freedom, who else will?

Notes

1 Isaiah Berlin, 'Two Concepts of Liberty,' in *Liberty*, by Isaiah Berlin, 2nd ed., Henry Hardy (Oxford: Oxford University Press, 2002), 166–217, https://doi.org/10.1093/019924989X.003.0004.

2 Ian Carter, 'Positive and Negative Liberty,' *Stanford Encyclopedia of Philosophy* 363, no. 9410 (2008): 1, https://doi.org/10.1016/S0140-6736(04)15657-2.

3 Carter, 7.

4 Philip Pettit, *Republicanism: A Theory of Freedom and Government*, 1, Issued in paperback, Oxford Political Theory (Oxford: Oxford University Press, 1999).

5 Pettit.

6 Carter, 'Positive and Negative Liberty,' 13.

7 Philip Pettit, 'Freedom as Antipower,' *Ethics* 106, no. 3 (April 1996): 578, https://doi.org/10.1086/233648.

8 A similar approach has been extensively explored by Marianna Capasso, 'Positions of Power. Freedom and Control in the Age of Artificial Intelligence' (Pisa, School of Advanced Studies Sant'Anna, 2022), https://dta.santannapisa.it/t/etd-12092021-161754/.

9 Langdon Winner, 'Do Artifacts Have Politics?,' *Daedalus* 109, no. 1 (1980): 121–36.

10 This is at least Winner's and others' interpretation. It is an open controversy whether this was the explicit intention of Moses. See Glenn Kessler, 'Robert Moses and the Saga of the Racist Parkway Bridges,' *Washington Post*, 11 November 2021, https://www.washingtonpost.com/politics/2021/11/10/robert-moses-saga-racist-parkway-bridges/.

11 Kate Crawford, *Atlas of AI: Power, Politics, and the Planetary Costs of Artificial Intelligence* (New Haven: Yale University Press, 2021).

12 Jack Stilgoe, 'Machine Learning, Social Learning and the Governance of Self-Driving Cars,' *Social Studies of Science* 48, no. 1 (2017): 25–56, https://doi.org/10.1177/0306312717741687; J. M. Bradshaw et al., 'The Seven Deadly Myths of "Autonomous Systems,"' *IEEE Intelligent Systems* 28, no. 3 (May 2013): 54–61, https://doi.org/10.1109/MIS.2013.70.

13 Bill Joy, 'Why the Future Doesn't Need Us,' *Wired*, January 2000, https://www.wired.com/2000/04/joy-2/.

14 Karl Marx, *Karl Marx: Selected Writings*, 2nd ed., repr (Oxford: Oxford University Press, 2000), 158.

15 Jim Dator, 'What Futures Studies Is, and Is Not,' in *Jim Dator: A Noticer in Time: Selected Work, 1967–2018*, ed. Jim Dator, Anticipation Science (Cham: Springer International Publishing, 2019), 3–5, https://doi.org/10.1007/978-3-030-17387-6_1.

16 Jack Stilgoe, Richard Owen, and Phil Macnaghten, 'Developing a Framework for Responsible Innovation,' *Research Policy* 42, no. 9 (2013): 1568–80, https://doi.org/10.1016/j.respol.2013.05.008; Richard Owen, René von Schomberg, and Phil Macnaghten, 'An Unfinished Journey? Reflections on a Decade of Responsible Research and Innovation,' *Journal of Responsible Innovation* 8, no. 2 (2021): 217–33, https://doi.org/10.1080/23299460.2021.1948789.

17 Batya Friedman, 'Value-Sensitive Design,' *Interactions* 3, no. 6 (December 1996): 16–23, https://doi.org/10.1145/242485.242493; Jeroen Van Den Hoven, Pieter E. Vermaas, and Ibo van de Poel, *Handbook of Ethics, Values, and Technological Design: Sources, Theory, Values and Application Domains* (Dordrecht: Springer, 2015), https://doi.org/10.1007/978-94-007-6970-0.

18 Anthony Dunne and Fiona Raby, *Speculative Everything: Design, Fiction, and Social Dreaming* (Cambridge, MA and London: The MIT Press, 2013).

19 Carl DiSalvo, *Design as Democratic Inquiry: Putting Experimental Civics into Practice* (Cambridge, MA: The MIT Press, 2022); Sasha Costanza-Chock, *Design Justice: Community-Led Practices to Build the Worlds We Need* (Cambridge, MA: The MIT Press, 2020).

20 James Muldoon and Paul Raekstad, 'Algorithmic Domination in the Gig Economy,' *European Journal of Political Theory* 22, no. 4 (2022): 587–607. https://doi.org/10.1177/14748851221082078; Marianna Capasso, 'Artificial Intelligence and the Future of Work. On the Emergence of Vulnerability Classes' (Paper presented at the International Conference on Structural Domination, TU Dresden, 18–19 Sept 2023).

21 Katherine C. Kellogg, Melissa A. Valentine, and Angéle Christin, 'Algorithms at Work: The New Contested Terrain of Control,' *Academy of Management Annals* 14, no. 1 (January 2020): 366–410, https://doi.org/10.5465/annals.2018.0174.

22 Tim Christiaens, *Digital Working Lives: Worker Autonomy and the Gig Economy* (Lanham: Rowman & Littlefield, 2023).

23 Carlo Vercellone, 'The Crisis of the Law of Value and the Becoming-Rent of Profit: Notes on the Systemic Crisis of Cognitive Capitalism,' in *Crisis in the Global Economy: Financial Markets, Social Struggles, and New Political Scenarios*, eds. Andrea Fumagalli and Sandro Mezzadra (Cambridge, MA: The MIT Press, 2010), 85–118.

24 Christian Fuchs and Sebastian Sevignani, 'What Is Digital Labour? What Is Digital Work? What's Their Difference? And Why Do These Questions Matter for Understanding Social Media?,' *TripleC* 11, no. 2 (6 June 2013): 237–93, https://doi.org/10.31269/triplec.v11i2.461.

25 Yann Moulier Boutang, *Cognitive Capitalism* (Cambridge: Polity Press, 2011).

26 Anca Gheaus and Lisa Herzog, 'The Goods of Work (Other than Money!),' *Journal of Social Philosophy* 47, no. 1 (1 March 2016): 70–89, https://doi.org/10.1111/josp.12140.

27 Michael Walzer, *Spheres of Justice: A Defense of Pluralism and Equality* (New York: Basic Books, 1983).

28 Filippo Santoni de Sio and Jeroen van den Hoven, 'Meaningful Human Control over Autonomous Systems: A Philosophical Account,' *Frontiers in Robotics and AI* 5 (February 2018): 15, https://doi.org/10.3389/frobt.2018.00015.

29 Filippo Santoni de Sio and Giulio Mecacci, 'Four Responsibility Gaps with Artificial Intelligence: Why They Matter and How to Address Them,' *Philosophy*

& Technology 34, no. 4 (14 December 2021): 1057–84, https://doi.org/10.1007/s13347-021-00450-x.

30 Miranda Fricker, *Epistemic Injustice: Power and the Ethics of Knowing* (Oxford: Oxford University Press, 2007).

31 Julie Simon et al., 'Digital Democracy: The Tools Transforming Political Engagement' (Nesta, 2017), https://www.nesta.org.uk/report/digital-democracy-the-tools-transforming-political-engagement/.

32 John Dewey, 'Challenge to Liberal Thought,' in *John Dewey: The Later Works, 1925–1953, Vol. 15, 1942–1948* (Carbondale, IL: Southern Illinois University Press, 1988), 261–75; Jane Addams, *Twenty Years at Hull-House with Autobiographical Notes* (New York: The MacMillan Company, 1910).

33 DiSalvo, 'Design as Democratic Inquiry,' 103–38.

34 Iris Marion Young, *Inclusion and Democracy* (Oxford: Oxford University Press, 2000).

35 Samuel Bagg, 'Sortition as Anti-Corruption: Popular Oversight against Elite Capture,' *American Journal of Political Science* 68, no. 1 (January 2024): 93–105, https://doi.org/10.1111/AJPS.12704.

36 Pettit, 'Freedom as Antipower.'

37 Elizabeth Tunstall, *Decolonizing Design: A Cultural Justice Guidebook* (Cambridge, MA: The MIT Press, 2023).

38 'Ecuador First to Grant Nature Constitutional Rights,' *Capitalism Nature Socialism* 19, no. 4 (2008): 131–33, https://doi.org/10.1080/10455750802575828.

39 Laura Forlano, 'Decentering the Human in the Design of Collaborative Cities,' *Design Issues* 32, no. 3 (1 July 2016): 42–54, https://doi.org/10.1162/DESI_a_00398.

40 Fleur Jongepier and Michael Klenk, eds., *The Philosophy of Online Manipulation* (New York: Routledge, 2022), https://doi.org/10.4324/9781003205425.

41 Van den Hoven, Jeroen et al., 'Privacy and Information Technology,' in *The Stanford Encyclopedia of Philosophy*, ed. Edward N. Zalta (Metaphysics Research Lab, Stanford University, 2020), https://plato.stanford.edu/archives/sum2020/entries/it-privacy/.

42 Shoshana Zuboff, *The Age of Surveillance Capitalism: The Fight for a Human Future at the New Frontier of Power* (London: Profile Books, 2019); Maša Galič, Tjerk Timan, and Bert-Jaap Koops, 'Bentham, Deleuze and Beyond: An Overview of Surveillance Theories from the Panopticon to Participation,' *Philosophy & Technology* 30, no. 1 (13 March 2017): 9–37, https://doi.org/10.1007/s13347-016-0219-1.

43 Christopher Burr and Luciano Floridi, eds., *Ethics of Digital Well-Being: A Multidisciplinary Approach*, vol. 140, Philosophical Studies Series (Cham: Springer International Publishing, 2020), https://doi.org/10.1007/978-3-030-50585-1.

44 Pratyusha Kalluri, 'Don't Ask If Artificial Intelligence Is Good or Fair, Ask How It Shifts Power,' *Nature* 583, no. 7815 (1 July 2020): 169, https://doi.org/10.1038/D41586-020-02003-2.

45 'UN AI Actions,' *AI for Good*, accessed 8 October 2023, https://aiforgood.itu.int/about-ai-for-good/un-ai-actions/.

46 'ACM FAccT Conference,' accessed 8 October 2023, https://facctconference.org/.

47 Thomas Metzinger, 'EU Guidelines: Ethics Washing Made in Europe,' *Tagesspiegel*, 4 August 2019, https://www.tagesspiegel.de/politik/eu-guidelines-ethics-washing-made-in-europe/24195496.html; Ben Wagner, 'Ethics As An

Escape From Regulation. From "Ethics-Washing" To Ethics-Shopping?,' in *Being Profiled*, eds. Emre Bayamlioglu et al. (Amsterdam: Amsterdam University Press, 2019), 84–9, https://doi.org/10.1515/9789048550180-016.

48 Annette Zimmermann, Kate Vredenburgh, and Seth Lazar, 'The Political Philosophy of Data and AI,' *Canadian Journal of Philosophy* 52, no. 1 (2022): 1–5, https://doi.org/10.1017/can.2022.28; Mark Coeckelbergh, *The Political Philosophy of AI: An Introduction* (Cambridge: Polity, 2022).

49 Virginia Eubanks, *Automating Inequality: How High-Tech Tools Profile, Police, and Punish the Poor* (New York: St. Martin's Press, 2018); Safiya Umoja Noble, *Algorithms of Oppression: How Search Engines Reinforce Racism* (New York: New York University Press, 2018).

50 DiSalvo, 'Design as Democratic Inquiry,' 30.

References

'ACM FAccT Conference.' Accessed 8 October 2023. https://facctconference.org/.

Addams, Jane. *Twenty Years at Hull-House with Autobiographical Notes*. New York: The MacMillan Company, 1910.

AI for Good. 'UN AI Actions.' Accessed 8 October 2023. https://aiforgood.itu.int/about-ai-for-good/un-ai-actions/.

Bagg, Samuel. 'Sortition as Anti-Corruption: Popular Oversight against Elite Capture.' *American Journal of Political Science* 68, no. 1 (January 2024): 93–105. https://doi.org/10.1111/AJPS.12704.

Berlin, Isaiah. 'Two Concepts of Liberty.' In *Liberty*, 2nd ed., edited by Isaiah Berlin and Henry Hardy, 166–217. Oxford: Oxford University Press, 2002. https://doi.org/10.1093/019924989X.003.0004.

Bradshaw, J. M., R. R. Hoffman, M. Johnson, and D. D. Woods. 'The Seven Deadly Myths of "Autonomous Systems".' *IEEE Intelligent Systems* 28, no. 3 (May 2013): 54–61. https://doi.org/10.1109/MIS.2013.70.

Burr, Christopher, and Luciano Floridi, eds. *Ethics of Digital Well-Being: A Multidisciplinary Approach*, Vol. 140. Philosophical Studies Series. Cham: Springer International Publishing, 2020. https://doi.org/10.1007/978-3-030-50585-1.

Capasso, Marianna. *Artificial Intelligence and the Future of Work. On the Emergence of Vulnerability Classes*. Paper presented at the International Conference on Structural Domination, TU Dresden, 18–19 Sept 2023.

———. 'Positions of Power. Freedom and Control in the Age of Artificial Intelligence.' *School of Advanced Studies Sant'Anna* (2022). https://dta.santannapisa.it/t/etd-12092021-161754/.

Carter, Ian. 'Positive and Negative Liberty.' *Stanford Encyclopedia of Philosophy* 363, no. 9410 (2008): 1–12. https://doi.org/10.1016/S0140-6736(04)15657-2.

Christiaens, Tim. *Digital Working Lives: Worker Autonomy and the Gig Economy. Off the Fence: Morality, Politics and Society*. Lanham: Rowman & Littlefield, 2023.

Coeckelbergh, Mark. *The Political Philosophy of AI: An Introduction*. Cambridge: Polity Press, 2022.

Costanza-Chock, Sasha. *Design Justice: Community-Led Practices to Build the Worlds We Need*. Cambridge, MA: The MIT Press, 2020.

Crawford, Kate. *Atlas of AI: Power, Politics, and the Planetary Costs of Artificial Intelligence*. New Haven: Yale University Press, 2021.

Dator, Jim. 'What Futures Studies Is, and Is Not.' In *Jim Dator: A Noticer in Time: Selected Work, 1967–2018*, edited by Jim Dator, 3–5. Anticipation Science. Cham: Springer International Publishing, 2019. https://doi.org/10.1007/978-3-030-17387-6_1.

Dewey, John. 'Challenge to Liberal Thought.' In *John Dewey: The Later Works, 1925–1953, Vol. 15, 1942–1948*, 261–75. Carbondale and Edwardsville: Southern Illinois University Press, 1988.

DiSalvo, Carl. *Design as Democratic Inquiry: Putting Experimental Civics into Practice*. Cambridge, MA: The MIT Press, 2022.

Dunne, Anthony, and Fiona Raby. *Speculative Everything: Design, Fiction, and Social Dreaming*. Cambridge, MA and London: The MIT Press, 2013.

'Ecuador First to Grant Nature Constitutional Rights.' *Capitalism Nature Socialism* 19, no. 4 (2008): 131–33. https://doi.org/10.1080/10455750802575828.

Eubanks, Virginia. *Automating Inequality: How High-Tech Tools Profile, Police, and Punish the Poor*. New York: St. Martin's Press, 2018.

Forlano, Laura. 'Decentering the Human in the Design of Collaborative Cities.' *Design Issues* 32, no. 3 (1 July 2016): 42–54. https://doi.org/10.1162/DESI_a_00398.

Fricker, Miranda. *Epistemic Injustice: Power and the Ethics of Knowing*. Oxford: Oxford University Press, 2007.

Friedman, Batya. 'Value-Sensitive Design.' *Interactions* 3, no. 6 (December 1996): 16–23. https://doi.org/10.1145/242485.242493.

Fuchs, Christian, and Sebastian Sevignani. 'What Is Digital Labour? What Is Digital Work? What's Their Difference? And Why Do These Questions Matter for Understanding Social Media?' *TripleC* 11, no. 2 (6 June 2013): 237–93. https://doi.org/10.31269/triplec.v11i2.461.

Galič, Maša, Tjerk Timan, and Bert-Jaap Koops. 'Bentham, Deleuze and Beyond: An Overview of Surveillance Theories from the Panopticon to Participation.' *Philosophy & Technology* 30, no. 1 (13 March 2017): 9–37. https://doi.org/10.1007/s13347-016-0219-1.

Gheaus, Anca, and Lisa Herzog. 'The Goods of Work (Other than Money!).' *Journal of Social Philosophy* 47, no. 1 (1 March 2016): 70–89. https://doi.org/10.1111/josp.12140.

Joy, Bill. 'Why the Future Doesn't Need Us.' *Wired*, January 2000. https://www.wired.com/2000/04/joy-2/.

Kalluri, Pratyusha. 'Don't Ask If Artificial Intelligence Is Good or Fair, Ask How It Shifts Power.' *Nature* 583, no. 7815 (1 July 2020): 169. https://doi.org/10.1038/D41586-020-02003-2.

Kellogg, Katherine C., Melissa A. Valentine, and Angéle Christin. 'Algorithms at Work: The New Contested Terrain of Control.' *Academy of Management Annals* 14, no. 1 (January 2020): 366–410. https://doi.org/10.5465/annals.2018.0174.

Kessler, Glenn. 'Robert Moses and the Saga of the Racist Parkway Bridges.' *Washington Post*, 11 November 2021. https://www.washingtonpost.com/politics/2021/11/10/robert-moses-saga-racist-parkway-bridges/.

Klenk, Michael, and Fleur Jongepier, eds. *The Philosophy of Online Manipulation*. New York: Routledge, 2022. https://doi.org/10.4324/9781003205425.

Marx, Karl. *Karl Marx: Selected Writings*, 2nd ed., Repr. Oxford: Oxford University Press, 2000.

Metzinger, Thomas. 'EU Guidelines: Ethics Washing Made in Europe.' *Tagesspiegel*, 4 August 2019. https://www.tagesspiegel.de/politik/eu-guidelines-ethics-washing-made-in-europe/24195496.html.

Moulier Boutang, Yann. *Cognitive Capitalism*. Cambridge: Polity Press, 2011.

Muldoon, James, and Paul Raekstad. 'Algorithmic Domination in the Gig Economy.' *European Journal of Political Theory* 22, no. 4 (2022): 587–607. https://doi.org/10.1177/14748851221082078.

Noble, Safiya Umoja. *Algorithms of Oppression: How Search Engines Reinforce Racism*. New York: New York University Press, 2018.

Owen, Richard, René von Schomberg, and Phil Macnaghten. 'An Unfinished Journey? Reflections on a Decade of Responsible Research and Innovation.' *Journal of Responsible Innovation* 8, no. 2 (2021): 217–33. https://doi.org/10.1080/23299460.2021.1948789.

Pettit, Philip. 'Freedom as Antipower.' *Ethics* 106, no. 3 (April 1996): 576–604. https://doi.org/10.1086/233648.

———. *Republicanism: A Theory of Freedom and Government*. Oxford: Oxford University Press, 1999.

Santoni de Sio, Filippo, and Giulio Mecacci. 'Four Responsibility Gaps with Artificial Intelligence: Why They Matter and How to Address Them.' *Philosophy & Technology* 34, no. 4 (14 December 2021): 1057–84. https://doi.org/10.1007/s13347-021-00450-x.

Santoni de Sio, Filippo, and Jeroen van den Hoven. 'Meaningful Human Control over Autonomous Systems: A Philosophical Account.' *Frontiers in Robotics and AI* 5 (February 2018): 15. https://doi.org/10.3389/frobt.2018.00015.

Simon, Julie, Theo Bass, Victoria Boelman, and Geoff Mulgan. *Digital Democracy: The Tools Transforming Political Engagement*. Nesta, 2017. https://www.nesta.org.uk/report/digital-democracy-the-tools-transforming-political-engagement/.

Stilgoe, Jack. 'Machine Learning, Social Learning and the Governance of Self-Driving Cars.' *Social Studies of Science* 48, no. 1 (2017): 25–56. https://doi.org/10.1177/0306312717741687.

Stilgoe, Jack, Richard Owen, and Phil Macnaghten. 'Developing a Framework for Responsible Innovation.' *Research Policy* 42, no. 9 (2013): 1568–80. https://doi.org/10.1016/j.respol.2013.05.008.

Tunstall, Elizabeth. *Decolonizing Design: A Cultural Justice Guidebook*. Cambridge, MA: The MIT Press, 2023.

Van den Hoven, Jeroen, Martijn Blaauw, Wolter Pieters, and Martijn Warnier. 'Privacy and Information Technology.' In *The Stanford Encyclopedia of Philosophy*, edited by Edward N. Zalta. Metaphysics Research Lab, Stanford University, 2020. https://plato.stanford.edu/archives/sum2020/entries/it-privacy/.

Van den Hoven, Jeroen, Pieter E. Vermaas, and Ibo van de Poel. *Handbook of Ethics, Values, and Technological Design: Sources, Theory, Values and*

Application Domains. Dordrecht: Springer, 2015. https://doi.org/10.1007/978-94-007-6970-0.

Vercellone, Carlo. 'The Crisis of the Law of Value and the Becoming-Rent of Profit: Notes on the Systemic Crisis of Cognitive Capitalism.' In *Crisis in the Global Economy: Financial Markets, Social Struggles, and New Political Scenarios*, edited by Andrea Fumagalli and Sandro Mezzadra, 85–118. Cambridge, MA: The MIT Press, 2010. https://api.semanticscholar.org/CorpusID:155386615.

Wagner, Ben. 'Ethics as an Escape from Regulation: From "Ethics-Washing" to Ethics-Shopping?' In *Being Profiled*, edited by Emre Bayamlioglu, Irina Baraliuc, Liisa Albertha Wilhelmina Janssens, and Mireille Hildebrandt, 84–89. Amsterdam University Press, 2019. https://doi.org/10.1515/9789048550180-016.

Walzer, Michael. *Spheres of Justice: A Defense of Pluralism and Equality*. New York: Basic Books1983.

Winner, Langdon. 'Do Artifacts Have Politics?' *Daedalus* 109, no. 1 (1980): 121–36.

Young, Iris Marion. *Inclusion and Democracy*. Oxford: Oxford University Press, 2000.

Zimmermann, Annette, Kate Vredenburgh, and Seth Lazar. 'The Political Philosophy of Data and AI.' *Canadian Journal of Philosophy* 52, no. 1 (2022): 1–5. https://doi.org/10.1017/can.2022.28.

Zuboff, Shoshana. *The Age of Surveillance Capitalism: The Fight for a Human Future at the New Frontier of Power*. London: Profile Books, 2019.

Part 1

AI, freedom, and design

2 AI, freedom, and the limits of Universal Basic Income

2.1 Artificial intelligence and Universal Basic Income

In November 2022, OpenAI launched ChatGPT, a chatbot that can automatically produce complex texts in a few seconds in reply to users' requests. Users may choose a specific language, format, style, and other aspects. Launched initially as a free service, ChatGPT has gained immense popularity among users (it attracted 100 million users in the first two months after its launch) and great attention from the media and the public. Looking at the relatively good quality of the performance of ChatGPT and the speed at which it has evolved from previous versions, many started wondering how long it will take for ChatGPT and similar applications to fully replace human professionals in activities such as drafting contracts, writing essays, journalistic pieces or technical codes, giving professional advice, and more.

ChatGPT is one form of artificial intelligence (henceforth AI). This term covers technologies that aim to replicate human intelligence and behaviour and, more generally, solve problems with relatively low direct control from human controllers. Specifically, ChatGPT is an AI based on the so-called large language model (LLM), a system that collects and processes vast amounts of text data, mainly from the internet. AI has recently been tested and used to assist and sometimes replace people in many domains, including medical diagnosing, public administration procedures, car driving, job selection procedures, military operations, and management of work activities via digital platforms such as Uber and others.

Whereas the explosion of ChatGPT has been, for many laypersons, a wake-up call about the potential of AI in assisting and replacing humans in complex tasks, this potential was well-known to experts. Popular books like Martin Ford's *The Rise of the Robots and the Threat of a Jobless Future*[1] and Brynjolfsson and MacAfee's *The Second Machine Age: Work, Progress, and Prosperity in a Time of Brilliant Technologies*[2] have claimed that AI may bring about a different type of change than previous technological and industrial revolutions, one that can potentially threaten industry and work as we know them. As illustrated by ChatGPT, AI allows the

DOI: 10.4324/9781003303244-3

automation of not only repetitive physical tasks but potentially any task that can be decomposed, including more complex tasks like driving as well as many mental activities that had long been thought to be impossible to automate, like accounting, planning, doing legal research, or medical diagnosing.[3] Moreover, from a broader economic perspective, the digital market has consistently taken a 'winner takes all' structure.[4] Companies that gain a significant market advantage in the digital domain tend not only to become wealthier than others but also to push others out of business, thereby further reducing employment opportunities. Think of Microsoft and Apple for computer software, Meta (Facebook and Instagram) and Twitter/X for social network platforms, and Google for search engines and other digital services (email, mapping, and others). All these might eventually bring industrialised societies towards what John Maynard Keynes in the 1930s called 'technological unemployment': 'unemployment due to our discovery of means of economising the use of labour outrunning the pace at which we can find new uses for labour.'[5]

Not so fast, insist some economists; let us not fall into the trap of what they call the 'Luddite fallacy.' The Luddites were a group of workers concerned about their jobs being replaced by machines in Great Britain during the First Industrial Revolution. They would organise protests against the introduction of machines on the work floor and even tried to destroy the machines. In the stories of many economists, Luddites play the role of the naïve, if not stupid, laypersons, unreasonably concerned about the possible negative impact of technological innovation on their lives and jobs, and unable to see the bigger picture, in which technology would eventually bring well-being for everybody. According to this story, the Luddite's reasoning – whether applied to old mechanical machines or to AI-driven robots – would be fallacious. Machines, no matter how sophisticated they may be, will make some jobs obsolete but will always create new jobs, and history has shown that in the long run, more jobs will be created than eliminated, and the job market will reach a new, better equilibrium. Some experts suggest that the AI revolution may not play out like older industrial revolutions: this time, jobs might disappear for good. In any case, the Luddites' concern was partly different: paraphrasing Keynes, they were concerned that in the long run, they might all be dead. In other words, they might have difficulties earning a living during the transition to a new paid occupation, assuming they could find one.[6]

In response to a possible increase in technological unemployment, a specific line of thinking has gained traction in the philosophical, political, and public debate on the future of work: responding to technological innovations with new, more robust policies for redistributing income, explicitly focusing on the impact of AI and robotics. One recurring theme, supported in the public debate by distant figures such as Microsoft's Bill Gates and

French Socialist Party representative Benoit Hamon, is the idea of a robot tax, to be paid by entrepreneurs investing in robotics, to compensate those who lose their jobs due to automation.[7] However, the most popular and controversial measure is the introduction of Universal Basic Income,[8] a policy that periodically provides each citizen with a fixed amount of money, irrespective of their income from other sources or willingness to work.

This chapter addresses the question of whether distributive policies like Universal Basic Income are sufficient to address the impact of AI on employment. In the first part of this chapter, I present two freedom-based arguments in support of Universal Basic Income: first, it would allow people to achieve 'real freedom' by giving them the possibility to pursue their preferred life path without depending on work (the real-libertarian argument); and, second, it would support freedom as non-domination by giving people the opportunity to resist dominating relationships at work through financial independence (the neo-republican argument).

In the second part of the chapter I propose three reasons to be sceptical about Universal Basic Income and other distributive measures being sufficient to protect people's freedom against the impact of AI on work: they fail to address losses of freedom coming with unemployment; they do not sufficiently address power relations in the workplace; they may (implicitly) support technological determinism.

2.2 The promises of Universal Basic Income

Universal Basic Income[9] has a long history that precedes the invention of digital technologies. It has been argued, in different forms, by thinkers with very different ideological profiles, such as the American founding father and revolutionary Thomas Paine, the civil rights activist Martin Luther King, and the Nobel laureate economist Milton Friedman. The concept entered the contemporary political debate with the work of the Belgian philosopher Philippe Van Parijs, synthesised in his book *Real Freedom for All*.[10] Van Parijs describes himself as a 'real libertarian.' He believes people can be free only if they have unconditional access to the resources needed to pursue their concept of a good life. Thus, he proposes the provision of a 'Universal Basic Income at the highest sustainable level.'[11] It is a pretty straightforward idea: you cannot be free if you cannot make your living. After Van Parijs, the idea of Universal Basic Income continued to be taken up by many different, sometimes conflicting, ideological perspectives, such as a feminist, [12] a socialist, [13] and a libertarian perspective. [14]

One alternative philosophical interpretation of Universal Basic Income is offered by neo-republican thinkers.[15] Neo-republicans reject the classical liberal notion of freedom, which holds that people are free so long as their choices are not interfered with by other agents. Neo-republicanism tries to take seriously the impact of power relations on freedom. According to the neo-republican notion of freedom, also called 'non-domination,' people are free so long as no agent is in a position of power to interfere with their choices arbitrarily. Thus, whereas traditional liberals identify unfreedom as *actual* choice interference, neo-republicans identify unfreedom (or domination) as *potential* arbitrary choice interference. You cannot be free if someone has the power to arbitrarily interfere with your choice, whether they exercise that power or not. As for work, some neo-republicans believe that Universal Basic Income may free up people in a specific way: by mitigating the power imbalance between employers and workers and making workers less dependent on the employer's arbitrary will. The financial resources provided by Universal Basic Income would give workers a real ability, as opposed to a merely formal right, to exit or avoid entering jobs that are exploitative or otherwise undesirable.[16] If you have some cash granted anyway, you are less dependent on the employer's will: you will be freer to speak up and leave, or so the argument goes.

So, according to two popular philosophical arguments, Universal Basic Income enhances people's freedom. It allows them to pursue their preferred concept of a good life without being concerned about making their living (Van Parijs' point). It ensures that workers are not being dominated by their employers thanks to their financial independence, giving them more power to change or leave their workplace if they do not like it (the neo-republican point).

At this point, it is important to note that the Universal Basic Income strategy is a post hoc or redistributive policy: it aims to distribute resources and power after the jobs have already been allocated. John Rawls, on whose work Philippe Van Parijs, the father of the Universal Basic Income movement, explicitly relies, offered some good reasons to be cautious about redistributive models such as Universal Basic Income being sufficient for social justice. In *Justice as Fairness*, echoing Marx, Rawls criticises a social structure which allows a small group of people to have a 'near monopoly' of productive assets. However, unlike Marx, he means not only the material means of production but also human capital: education and trained skills. Therefore, Rawls proposes that society is organised to ensure a 'widespread ownership of productive assets and human capital.'[17] This is one key feature of what he calls 'property-owning democracy.' In a property-owning democracy, the 'narrowing and demeaning features' of the capitalist division of labour would be 'largely overcome.'[18] Through access to education and trained skills, people could avoid demeaning jobs and tasks.

Universal Basic Income's supporters are undoubtedly correct in stressing the importance of financial independence to protect people from poverty and coercive, exploitative, or dominating (work) relationships.[19] And a better distribution of education and skills invoked by Rawls and others is a valuable measure to pursue. However, in the following sections of this chapter, I propose some criticisms of these measures to protect and promote human freedom in relation to new technologies in the workplace.

2.3 Limits of redistributive policies such as Universal Basic Income

First, I want to make clear what my criticisms are *not* about. There are two standard lines of complaints towards distributive practices like Universal Basic Income. The most substantial criticism holds that Universal Basic Income is unfair because it would allow or even encourage people to live on the work of others. [20] This is not my criticism, and I here agree with the point iconically made by Karl Widerquist in his 2006 essay 'Who Exploits Who?'[21]: even accepting that, under Universal Basic Income, some people may somehow take advantage of the work of others (by deciding not to work and benefitting from the free income granted by the Universal Basic Income), this would be a minor and less serious problem compared to what happens in the current (welfare) systems, where people are forced to accept exploitative and unfair working conditions to keep their job or to maintain their unemployment benefits. Another criticism considers Universal Basic Income politically or economically unfeasible.[22] While I share this view, this is not my main concern.

My concern here is that Universal Basic Income and property-owning democracy, while being potentially effective in alleviating some forms of unfreedom and injustice, are not sufficient to achieve the goal of enhancing human freedom in the face of the current technological developments. Therefore, they must be complemented with other practices and interventions including new ways of designing technology at work.

In the rest of the chapter, I will first present three limits of distributive policies in addressing the impact of technology on the workplace. Distributive policies:

(1) Do not address the restriction of (work) options (due to technological development) and the disempowerment of people due to the weakening of social scaffolding offered by work
(2) Do not sufficiently consider threats to freedom coming from power relations in the workplace (supported by new technologies)
(3) May indirectly support a passive approach towards technological development (so-called technological determinism)

In Chapter 3, I will then present the potential of a more proactive, design-oriented attitude to support worker's freedom, one that complements more traditional policy interventions. According to this 'design stance,' we should address social issues related to technological development at their roots, that is, we should address technological development itself: rethinking the general processes and power dynamics behind it (who determines the direction of technological development and how) and the design process (who determines design choices and how). In a nutshell, I propose considering technological development and design as a political issue, the subject of social debate and proactive social and political intervention, not as an independent variable that society and politics can only passively look at and react to.

2.3.1 *First limit of distributive policies: they fail to address losses of freedom coming with unemployment*

At least in the eyes of those who propose it as an answer to the risks of technological unemployment caused by AI (e.g. Martin Ford[23] and Mark Walker[24]), Universal Basic Income is needed to compensate those whom the digital revolution will (permanently) leave without paid work. So, these basic income strategies are silent about possible plans to prevent this structural mass unemployment from occurring in the first place. This is a problem for 'real freedom.' Real freedom arguably requires not only that people are not obliged to do paid work they consider unfair or not valuable but also that they must be able to do work that they regard as valuable. Whereas 'the end of work' may be good news for those who see it mainly as a source of suffering and exploitation, [25] (paid) work is, for some, a source of many 'goods other than money'[26] such as acquisition of skills and self-identity, and engagement in social life. Moreover, one may want a paid job to avoid the stigma and discrimination associated with unemployment in an industrialised society[27] or for other personal reasons. In this perspective, Universal Basic Income risks becoming more of a monetary compensation for the loss of freedom of choice than a means to protect or increase it. Freedom *from* work, even if compensated by some money, is not the freedom everybody wants. Whereas Universal Basic Income may help those left unemployed, it does not seem sufficient to address and govern the impact on human freedom of introducing AI in the workplace. This implies the acceptance of technological unemployment as a given, an attitude which, ironically, may undermine the aims of Van Parijs' real libertarianism: keeping relevant options open for people.

However, some Universal Basic Income supporters believe that Universal Basic Income will make the persons left unemployed by technological change willing and able to engage in other (more) valuable forms of work activities outside the boundaries of traditional paid job contracts, such as care or social work, or cultural or meaningful leisure activities. But they do not say where they will find the many resources and how they will develop the skills and motivation required to do so. Universal Basic Income and other forms of financial support aim to provide people with the means to make a living, not the resources to develop the capacities and opportunities to engage in new social, intellectual, professional, leisure, or artistic activities. Many doubt that a Universal Basic Income might ever be high enough to grant workers real independence from the labour market,[28] so it sounds implausible to assume a Universal Basic Income sufficiently high to cover for investments in self-development, education, trainings, meetings, activities, travel, etc.

Finally, personal self-development requires material resources and a social environment that promotes and supports it.[29] It has recently been argued that freedom is 'relational' and 'socially scaffolded': it crucially depends on one's position in a system of relationships, norms, and practices,[30] and it requires the support of many formal and informal social and political structures.[31] While a little sum of cash can certainly be beneficial to make one's living, it's doubtful that it could be sufficient to create a free life without a supporting social environment. One may even argue that the significant financial commitment by the State to support Universal Basic Income would likely reduce the State's capacity to invest in promoting these structures. Universal Basic Income may not provide universal basic *freedom*.

In this sense, property-owning democracy – the idea of achieving a better distribution of human resources, typically skills and education needed to access a broader variety of job opportunities – has its limitations as well. The need for a (re)distribution of education and relevant skills is a big theme in the public debate on AI and the future of work. For instance, MIT's Brinyolfsson and McAfee (authors of the popular *The Second Machine Age* book) insist on rethinking education and skill formation as a key to flourishing in the new, AI-dominated society. They argue that with AI and robotics, many jobs and tasks will disappear but many others will remain, and those with the relevant skills will remain employed. So, in this perspective, programmes to provide people with the appropriate skills – for instance, via lifelong learning – are the way to go. However, redistributing skills and other human capital will not create more desirable jobs; 'too many skilled people would be competing for still limitedly available meaningful jobs – which would likely lead most of them to take precisely "narrowing and demeaning" work' that property-owning democracy is supposed to prevent.[32] Notwithstanding the recurring hope that future robots will take all the dull, dangerous, and dirty jobs or tasks, we don't seem close to this future.[33] These jobs will not disappear soon, and the many who will end up doing them are likely to be people in socially discriminated categories, according to what has been called 'occupational segregation.'[34] So, in other words, a better (pre-)distribution of material and social resources will not necessarily affect the distribution of labour. As with the mere possession of some financial help (Universal Basic Income), possessing human capital, education, and skills may not necessarily give (all) workers sufficient freedom and power to try to obtain the kind of work they want and need.[35]

While pointing to two very important, possibly necessary elements to promote human freedom and justice in the search for good work – financial independence and the possession of education and skill – distributive models such as Universal Basic Income and property-owning democracy cannot grant realisation of such freedom of choice for all.

2.3.2 *Second limit of distributive policies: they do not sufficiently address power relations in the workplace*

There is one more, possibly more important, issue for the scope of this book. Distributive justice approaches focus on equal opportunity to choose a job (property-owning democracy) or even to not get a traditional paid job (Universal Basic Income). But they tend to be silent about promoting freedom or justice *within work*. New technologies will not eliminate work any time soon, but they are very likely to change how work is done, reshaping tasks and relationships between workers and between workers and managers.[36] Distributive justice measures such as (pre)distribution of material and educational resources can do much but not everything. Recent studies in social science and political philosophy show that some specific forms of injustice rooted in work structures may not be affected by those measures. The way a company is organised, including the division of tasks, the quality of one's work, the social division of labour, and free time – are elements that do not seem to be reducible to issues of maldistribution of goods or resources. This concerns both the qualitative side of work (*what* people are requested or allowed to do and *how*) and the quantitative side (*when* and *how much*).[37] In other words, some specific forms of injustice, inequality, and oppression in the workspace may persist even in a society where distributive justice outside the workspace is largely achieved.

Social inequalities and oppression in the workspace may happen at the level of interaction between individuals, in the form of (dis)respectful interactions between individual managers and employees or individual employees among them. Or, at the structural level, in the form of attributing specific status to certain groups of people (as identified by gender, race, income, etc.), systematically affecting their capacity or opportunity to be assigned more interesting tasks, more convenient work times, and other perks. Notably, the interactional and structural dimensions are quite independent so that someone may enjoy respectful and caring individual relationships with managers and colleagues while at the same time being structurally disadvantaged. Or, conversely, someone not suffering any structural disadvantage may still be treated disrespectfully by individual colleagues and managers.[38]

Significantly, the quality of relations at work will dramatically affect the general quality of one's life. Harmful work interactions will most likely affect individual well-being outside the workspace and the possibility to pursue a good life. And structural discrimination in the workspace will create or reinforce social discrimination outside. When we combine this with the fact that social inequalities outside the workspace are likely to be reproduced in the workplace, as emphasised in the theories of distributive justice, we arrive at that so-called 'status circle', a loop of mutually reinforcing inequalities between the workplace and the outside world.[39]

From a philosophical point of view, injustice at work can be conceptualised as a problem for equality and freedom. As for equality, one can see this as a violation of the principle of the 'equal moral worth' of persons,[40] that of social equality[41] or that of 'participatory parity,' the equal capacity to participate in social life as peers.[42] As for freedom, the presence of social structures giving some groups arbitrary or uncontrolled power over others is a classic example of domination. And being dominated, following the neo-republican insight, is incompatible with the status of a free person. I will return to domination in the workplace, with a specific focus on the role of digital technologies, in Chapter 4.

As described above, some neo-republican supporters of Universal Basic Income rely on the idea that the financial independence granted by Universal Basic Income will give employees more power to speak up for themselves or walk away from dominating work relationships. But this is easier said than done. Workers often internalise the intimidation of their employers and may not be confident enough to speak up or decide to leave.[43] They may also reasonably believe that they are unlikely to find a new job that is not similarly dominating insofar as it is a ubiquitous practice among capitalist enterprises to give managers sufficient power to make their workers' lives miserable or even humiliate them. [44] Indeed, as we will see in the Chapter 4, technological development may make specific forms of control and surveillance readily available to many employers, so the freedom to leave one's current employer may not solve the problem.[45]

Distributive justice frameworks do not sufficiently address (new) power structures in the workplace. They are insufficient to support justice in the workplace and do not give people enough power to get out from dominating work relationships if the (new) structures of power are ubiquitous across workplaces.

2.3.3 *Third limit of distributive policies: they may support technological determinism*

I have argued above that one problem with distributive measures like Universal Basic Income is that they tend to consider the realisation of technological unemployment as a sort of independent variable, something that may or may not happen but that falls beyond the scope of social and political choices. Typically, distributive justice measures aim to protect and promote human freedom by counterbalancing the effects of technological development on employment. Similarly, they offer workers (financial) tools to be less exposed to the economic domination of employers and ways to develop skills and capabilities to be more able to avoid or resist coercive, demeaning, or otherwise undesirable work. My criticism has been that these measures, as commendable and effective as they may be in offering people financial support, are insufficient to fully support human freedom. This is because they do not address the causes of (undesired) technological unemployment nor the source of unfreedom and injustice in the workplace.

Taken together, these two limits constitute the third limit identified in this chapter: distributive justice measures bring the risk of implicitly accepting some form of so-called 'technological determinism,' that is, treating technological development as an autonomous and ungovernable force, or at least remaining silent about social and political interventions needed to promote freedom and justice *in* technological development and *in the design* and implementation of the technology that will affect work.

In his 1977 book *Autonomous Technology: Technique-Out-of-Control as a Theme in Political Thought*, the US philosopher and sociologist of technology Langdon Winner presented the idea of 'the technological society' as a recurrent theme in the philosophy of that time.[46] The French philosopher Jacques Ellul and others had presented technological thinking as a powerful cultural force shaping modern society in all its dimensions.[47] Some interpreted this in a deterministic way as an unstoppable trend or, in Winner's words, as a 'technique-out-of-control.' Others insisted that forms of critique, opposition, and reform become possible once this trend has been identified. Among the latter was arguably Winner himself, who proposed an intriguing concept at the end of his book: 'epistemological luddism'.

According to Winner, we should not passively accept the discourse on technique-out-of-control or what later has also been called – with more specific reference to AI and robotics – the 'myth of autonomous technology.'[48] This may turn into a self-fulfilling prophecy: the belief in technological determinism, as it were, makes technological determinism true. If we believe that technology is ungovernable, because we believe it and, therefore, we refrain from trying to govern it, then it will necessarily keep developing in uncontrollable directions. As the sociologist and philosopher of technology Jack Stilgoe said: 'Claims that technology has a will of its own … typically disguise a political agenda that is libertarian and deregulatory.'[49] At the same time, believing in human freedom to shape technological development in different possible directions opens the possibility of governing technological development. However, this is also easier said than done because taking such an 'attitude of freedom' towards technology requires us to 'put into question the human-built world into which we are born.'[50] This requires a strong epistemological and cultural change rather than just a political effort. We must first find ways to destroy our mental image of the (technological) world and its necessary constituents if we want to be able to rebuild a different world and a different technology. This will be one of the recurring themes of this book. Promoting freedom in a technological society requires asking ourselves two questions: how can we design technology that respects or promotes human freedom and what design practices encourage the capacity *to think freely* about technology?

Winner identifies a starting point for this epistemological effort in his later essay: 'Do Artifacts Have Politics?'[51] The first step to realise techno-societal change is the realisation that technology is never neutral since its design and implementation often reflect the values and principles of the actors who designed and conceived it. Technology always impacts the promotion of specific values and principles and the opportunities and powers of different societal actors. This is true of simple artefacts that have nothing to do with AI. The most famous example discussed by Winner is an overbridge designed by the New York urban planner Robert Moses. This relatively simple and apparently politically neutral piece of the built environment was, according to Winner, designed and used with the clear intention to achieve a (dubious) social and political goal: preventing public transport from reaching Long Island beach, thereby preventing poor (Black) people from getting there.[52] As it happens, technology's ethical, social, and political effects are not always predicted or explicitly intended by their designers, like in this case. They are often more complicated. Later on, I will introduce the so-called 'postphenomenological' approach to technology, according to which designers cannot determine the nature of their products once and for all, as objects and humans (users) constantly shape each other in their interactions.[53] Still, and even more so, these complex normative structures and political effects of technology should be critically and continuously analysed and made explicit if we want to effectively address and govern them.

In the next chapter and elsewhere in this book, I will claim that freedom and other values can be better protected by the systematic application of a *design stance* towards the development, governance, and use of AI and technology in general. According to such a design stance, no matter how formidable technology and the economic forces propelling it are we should avoid succumbing to an (implicit) technological determinism that obscures the institutional and socio-technical design space and discourages the proposal of new designs. Technological advances can be beneficial to society only if designers and policymakers are willing and able to proactively embed more relevant human and social values into the process of designing and developing the new technological systems, as opposed to just regulating their use or devising post hoc policies to govern their ethical and societal impact.[54]

It is often remarked that, whether we like it or not, to some extent and in some way, technology such as AI *will* be introduced in society. This is probably true. The question is, indeed, whether we like it or not, to what extent and in what way this should happen.

Notes

1 Martin Ford, *Rise of the Robots: Technology and the Threat of a Jobless Future* (New York: Basic Books, 2015).
2 Erik Brynjolfsson and Andrew MacAfee, *The Second Machine Age: Work, Progress, and Prosperity in a Time of Brilliant Technologies* (New York: Norton and Company, 2014).
3 Ford, *Rise of the Robots: Technology and the Threat of a Jobless Future.*
4 Robert H. Frank and Philip J. Cook, *The Winner-Take-All Society: Why the Few at the Top Get So Much More Than the Rest of Us* (London: Penguin Books, 1996).
5 John Maynard Keynes, 'Economic Possibilities for Our Grandchildren,' in John Maynard Keynes, *Essays in Persuasion* (New York: Harcourt Brace, 1930), 358–73.
6 Jelena Belic, 'Institutions, Automation, and Legitimate Expectations,' *The Journal of Ethics*, 12 September 2023, https://doi.org/10.1007/s10892-023-09440-x.
7 Ryan Abbott and Bret Bogenschneider, 'Should Robots Pay Taxes? Tax Policy in the Age of Automation,' *SSRN Electronic Journal*, 23 March 2018, https://doi.org/10.2139/ssrn.2932483.
8 Mark Walker, 'BIG and Technological Unemployment: Chicken Little Versus the Economists,' *Journal of Evolution and Technology* 24, no. 1 (2014): 5–25; Ford, *Rise of the Robots: Technology and the Threat of a Jobless Future.*
9 Credit is due to Txai Almeida for this section's philosophical discussion about Universal Basic Income, which is based on Filippo Santoni de Sio, Txai Almeida, and Jeroen van den Hoven, 'The Future of Work: Freedom, Justice and Capital in the Age of Artificial Intelligence,' *Critical Review of International Social and Political Philosophy*, 13 December 2021, 1–25, https://doi.org/10.1080/13698230.2021.2008204.
10 Philippe van Parijs, *Real Freedom for All: What (If Anything) Can Justify Capitalism?, Oxford Political Theory* (Oxford: Clarendon Press, 1995).
11 Van Parijs, 5.
12 Julieta M. Elgarte, 'Basic Income and the Gendered Division of Labour,' *Basic Income Studies* 3, no. 3 (18 April 2008), https://doi.org/10.2202/1932-0183.1136; Anne Alstott, 'Good for Women,' in *What's Wrong with a Free Lunch?*, eds. Joshua Cohen, Joel Rogers, and Philippe Van Parijs (Boston: Boston Beacon Press, 2001), 75–79.
13 Erik Olin Wright, 'Basic Income as a Socialist Project,' *Basic Income Studies* 1, no. 1 (4 March 2006), https://doi.org/10.2202/1932-0183.1008; David Calnitsky, 'Debating Basic Income,' *Catalyst* 1, no. 3 (2017): 62–91.
14 Matt Zwolinski, 'A Hayekian Case for Free Markets and a Basic Income,' in *The Future of Work, Technology, and Basic Income*, eds. Michael Cholbi and Michael Weber (London and New York: Routledge, 2019), 7–26.
15 Philip Pettit, 'A Republican Right to Basic Income?,' *Basic Income Studies* 2, no. 2 (23 October 2008), https://doi.org/10.2202/1932-0183.1082; David Casassas, 'Basic Income and the Republican Ideal: Rethinking Material Independence in Contemporary Societies,' *Basic Income Studies* 2, no. 2 (18 April 2008), https://doi.org/10.2202/1932-0183.1081.

16 Pettit, 'A Republican Right to Basic Income?'; Casassas, 'Basic Income and the Republican Ideal: Rethinking Material Independence in Contemporary Societies.'
17 John Rawls, *Justice as Fairness: A Restatement*, ed. Erin I. Kelly (Cambridge, MA: Belknap Press, 2001), 139.
18 Rawls, 177.
19 Karl Widerquist, 'Who Exploits Who?,' *Political Studies* 54, no. 3 (October 2006): 444–64, https://doi.org/10.1111/J.1467-9248.2006.00614.X.
20 Robert Van der Veen, 'Real Freedom versus Reciprocity: Competing Views on the Justice of Unconditional Basic Income,' *Political Studies* 40, no. 6 (1998): 140–63; G. Van Donselaar, *The Right to Exploit: Parasitism, Scarcity, Basic Income* (Oxford and New York: Oxford University Press, 2009).
21 Widerquist, 'Who Exploits Who?'
22 Alex Gourevitch and Lucas Stanczyk, 'The Basic Income Illusion,' *Catalyst* 1, no. 4 (2018): 151–177.
23 Ford, *Rise of the Robots: Technology and the Threat of a Jobless Future.*
24 Walker, 'BIG and Technological Unemployment: Chicken Little Versus the Economists.'
25 John Danaher, 'Will Life Be Worth Living in a World Without Work? Technological Unemployment and the Meaning of Life,' *Science and Engineering Ethics* 23, no. 1 (11 February 2017): 41–64, https://doi.org/10.1007/s11948-016-9770-5.
26 Anca Gheaus and Lisa Herzog, 'The Goods of Work (Other than Money!),' *Journal of Social Philosophy* 47, no. 1 (1 March 2016): 70–89, https://doi.org/10.1111/josp.12140.
27 Daniel Sage and Patrick Diamond, 'Europe's New Social Reality: The Case Against Universal Basic Income' (Policy Network, 2017), http://www.policy-network.net/publications/6190/Europes-New-Social-Reality-the-Case-Against-Universal-Basic-Income.
28 Simon Birnbaum and Jurgen De Wispelaere, 'Exit Strategy or Exit Trap? Basic Income and the "Power to Say No" in the Age of Precarious Employment,' *Socio-Economic Review*, 14 February 2020, mwaa002, https://doi.org/10.1093/ser/mwaa002.
29 Amartya Sen, *The Idea of Justice* (Cambridge, MA: Harvard University Press, 2009), https://www.hup.harvard.edu/catalog.php?isbn=9780674060470.
30 Catriona Mackenzie et al., eds., *Relational Autonomy: Feminist Perspectives on Autonomy, Agency, and the Social Self* (Oxford and New York: Oxford University Press, 2000).
31 Philip Pettit and Victoria McGeer, 'The Hard Problem of Responsibility,' in *Oxford Studies in Agency and Responsibility, 3*, ed. David Shoemaker (Oxford University Press, 2015), 160–188.
32 Denise Celentano, 'Work Justice Beyond Redistribution and Recognition,' in *Paradigms of Justice*, eds. Denise Celentano and Luigi Caranti (London and New York: Routledge, 2020), 90–117.
33 Leslie Willcocks, 'Robo-Apocalypse Cancelled? Reframing the Automation and Future of Work Debate,' *Journal of Information Technology* 35, no. 4 (10 December 2020): 286–302, https://doi.org/10.1177/0268396220925830.
34 Paul Gomberg, *How to Make Opportunity Equal: Race and Contributive Justice* (Malden, MA: Blackwell Pub, 2007).
35 For a broader discussion of this and other related points inspired by the work of authors such as Iris Young, Elizabeth Anderson and Paul Gomberg, see Celentano, 'Work Justice Beyond Redistribution and Recognition.'

36 Willcocks, 'Robo-Apocalypse Cancelled? Reframing the Automation and Future of Work Debate.'
37 Celentano, 'Work Justice Beyond Redistribution and Recognition,' 112.
38 Celentano, 106–7.
39 Cecilia L. Ridgeway, 'Why Status Matters for Inequality,' *American Sociological Review* 79, no. 1 (February 2014): 1–16, https://doi.org/10.1177/0003122413515997.
40 Elizabeth S. Anderson, 'What Is the Point of Equality?,' *Ethics* 109, no. 2 (18 November 1999): 287–337, https://doi.org/10.1086/233897.
41 David Miller, 'Equality and Justice,' *Ratio* 10, no. 3 (1997): 222–37, https://doi.org/10.1111/1467-9329.00042.
42 Nancy Fraser, 'Rethinking Recognition: Overcoming Displacement and Reification in Cultural Politics,' no. 3 (2000); Celentano, 'Work Justice Beyond Redistribution and Recognition,' 108.
43 Alex Gourevitch, 'Labor Republicanism and the Transformation of Work,' *Political Theory* 41, no. 4: 591–617, https://doi.org/10.1177/0090591713485370.
44 Elizabeth Anderson, *Private Government: How Employers Rule Our Lives (and Why We Don't Talk about It), University Center for Human Values Series* (Princeton and Oxford: Princeton University Press, 2017).
45 In Philip Pettit, *On the People's Terms: A Republican Theory and Model of Democracy* (Cambridge: Cambridge University Press, 2012), Pettit admits that Universal Basic Income may not be sufficient to grant workers the freedom to resist the employers' power and that other forms of protection and regulation of work are necessary.
46 Langdon Winner, *Autonomous Technology. Technics-out-of-Control as a Theme in Political Thought* (Cambridge, MA: The MIT Press, 1977).
47 Jacques Ellul, *The Technological Society: A Penetrating Analysis of Our Technical Civilization and of the Effect of an Increasingly Standardized Culture on the Future of Man* [Nachdruck der Ausgabe] (New York, NY: Knopf, Vintage Books, 2011).
48 J. M. Bradshaw et al., 'The Seven Deadly Myths of "Autonomous Systems,"' *IEEE Intelligent Systems* 28, no. 3 (May 2013): 54–61, https://doi.org/10.1109/MIS.2013.70; Jack Stilgoe, 'Machine Learning, Social Learning and the Governance of Self-Driving Cars,' *Social Studies of Science* 48, no. 1 (2017): 25–56, 11, https://doi.org/10.1177/0306312717741687.
49 Stilgoe, 'Machine Learning, Social Learning and the Governance of Self-Driving Cars,' 11.
50 Alfred Nordmann, 'Epistemological Luddism,' *Engaging Science, Technology, and Society* 7, no. 1 (5 October 2021): 109–12, https://doi.org/10.17351/ests2021.861.
51 Langdon Winner, 'Do Artifacts Have Politics?,' *Daedalus* 109, no. 1 (1980): 121–36.
52 This is at least Winner's interpretation. It is an open controversy whether this was the explicit intention of Moses. See Glenn Kessler, 'Robert Moses and the Saga of the Racist Parkway Bridges,' *Washington Post*, 11 November 2021, https://www.washingtonpost.com/politics/2021/11/10/robert-moses-saga-racist-parkway-bridges/.
53 Peter-Paul Verbeek, *What Things Do: Philosophical Reflections on Technology, Agency, and Design* (University Park, PA: Penn State University Press, 2005).
54 Van den Hoven, Jeroen, 'Value Sensitive Design and Responsible Innovation,' in *Responsible Innovation*, eds. Richard Owen, John Bessant, and Maggy Heintz (Chichester: John Wiley & Sons, Ltd, 2013), 75–83, https://doi.org/10.1002/9781118551424.ch4.

References

Abbott, Ryan, and Bret Bogenschneider. 'Should Robots Pay Taxes? Tax Policy in the Age of Automation.' *SSRN Electronic Journal* (23 March 2018). https://doi.org/10.2139/ssrn.2932483.

Alstott, Anne. 'Good for Women.' In *What's Wrong with a Free Lunch?*, edited by Joshua Cohen, Joel Rogers, and Philippe Van Parijs. Boston: Boston Beacon Press, 2001, 75–79.

Anderson, Elizabeth S. *Private Government: How Employers Rule Our Lives (and Why We Don't Talk about It)*. University Center for Human Values Series. Princeton and Oxford: Princeton University Press, 2017.

———. 'What Is the Point of Equality?' *Ethics* 109, no. 2 (18 November 1999): 287–337. https://doi.org/10.1086/233897.

Belic, Jelena. 'Institutions, Automation, and Legitimate Expectations.' *The Journal of Ethics* (12 September 2023). https://doi.org/10.1007/s10892-023-09440-x.

Birnbaum, Simon, and Jurgen De Wispelaere. 'Exit Strategy or Exit Trap? Basic Income and the "Power to Say No" in the Age of Precarious Employment.' *Socio-Economic Review* (14 February 2020). https://doi.org/10.1093/ser/mwaa002.

Bradshaw, J. M., R. R. Hoffman, M. Johnson, and D. D. Woods. 'The Seven Deadly Myths of "Autonomous Systems".' *IEEE Intelligent Systems* 28, no. 3 (May 2013): 54–61. https://doi.org/10.1109/MIS.2013.70.

Brynjolfsson, Erik, and Andrew MacAfee. *The Second Machine Age: Work, Progress, and Prosperity in a Time of Brilliant Technologies*. New York: Norton and Company, 2014.

Calnitsky, David. 'Debating Basic Income.' *Catalyst* 1, no. 3 (2017): 62–91.

Casassas, David. 'Basic Income and the Republican Ideal: Rethinking Material Independence in Contemporary Societies.' *Basic Income Studies* 2, no. 2 (18 April 2008). https://doi.org/10.2202/1932-0183.1081.

Celentano, Denise. 'Work Justice Beyond Redistribution and Recognition.' In *Paradigms of Justice: Redistribution, Recognition, and Beyond*, edited by Denise Celentano and Luigi Caranti. London and New York: Routledge, 2021, 90–117.

Danaher, John. 'Will Life Be Worth Living in a World Without Work? Technological Unemployment and the Meaning of Life.' *Science and Engineering Ethics* 23, no. 1 (11 February 2017): 41–64. https://doi.org/10.1007/s11948-016-9770-5.

Elgarte, Julieta M. 'Basic Income and the Gendered Division of Labour.' *Basic Income Studies* 3, no. 3 (18 April 2008). https://doi.org/10.2202/1932-0183.1136.

Ellul, Jacques. *The Technological Society: A Penetrating Analysis of Our Technical Civilization and of the Effect of an Increasingly Standardized Culture on the Future of Man*. New York: Vintage Books, 2011.

Ford, Martin. *Rise of the Robots: Technology and the Threat of a Jobless Future*. New York: Basic Books, 2015.

Frank, Robert H., and Philip J. Cook. *The Winner-Take-All Society: Why the Few at the Top Get So Much More Than the Rest of Us*. London: Penguin Books, 1996.

Fraser, Nancy. 'Rethinking Recognition: Overcoming Displacement and Reification in Cultural Politics.' *New Left Review* 3, no. 3 (2000): 107–120.

Gheaus, Anca, and Lisa Herzog. 'The Goods of Work (Other than Money!)' *Journal of Social Philosophy* 47, no. 1 (1 March 2016): 70–89. https://doi.org/10.1111/josp.12140.

Gomberg, Paul. *How to Make Opportunity Equal: Race and Contributive Justice*. Malden, MA: Blackwell Pub, 2007.

Gourevitch, Alex. 'Labor Republicanism and the Transformation of Work.' *Political Theory* 41, no. 4: 591–617. https://doi.org/10.1177/0090591713485370.

Gourevitch, Alex, and Lucas Stanczyk. 'The Basic Income Illusion.' *Catalyst* 1, no. 4 (2018): 151–177.

Kessler, Glenn. 'Robert Moses and the Saga of the Racist Parkway Bridges.' *Washington Post*, 11 November 2021. https://www.washingtonpost.com/politics/2021/11/10/robert-moses-saga-racist-parkway-bridges/.

Keynes, John Maynard. 'Economic Possibilities for Our Grandchildren.' In John Maynard Keynes, *Essays in Persuasion*, 358–73. New York: Harcourt Brace, 1930.

Mackenzie, Catriona, Natalie Stoljar, Catriona Mackenzie, and Natalie Stoljar, eds. *Relational Autonomy: Feminist Perspectives on Autonomy, Agency, and the Social Self*. Oxford and New York: Oxford University Press, 2000.

Miller, David. 'Equality and Justice.' *Ratio* 10, no. 3 (1997): 222–37. https://doi.org/10.1111/1467-9329.00042.

Nordmann, Alfred. 'Epistemological Luddism.' *Engaging Science, Technology, and Society* 7, no. 1 (5 October 2021): 109–12. https://doi.org/10.17351/ests2021.861.

Parijs, Philippe van. *Real Freedom for All: What (If Anything) Can Justify Capitalism? Oxford Political Theory*. Oxford: Clarendon Press, 1995.

Pettit, Philip. *On the People's Terms: A Republican Theory and Model of Democracy*. Cambridge: Cambridge University Press, 2012.

———. 'A Republican Right to Basic Income?' *Basic Income Studies* 2, no. 2 (23 October 2008). https://doi.org/10.2202/1932-0183.1082.

Pettit, Philip, and Victoria McGeer. 'The Hard Problem of Responsibility.' In *Oxford Studies in Agency and Responsibility, 3*, edited by David Shoemaker, 160–188. Oxford: Oxford University Press, 2015.

Rawls, John. *Justice as Fairness: A Restatement*. Edited by Erin I. Kelly. Cambridge, MA: Belknap Press, 2001.

Ridgeway, Cecilia L. 'Why Status Matters for Inequality.' *American Sociological Review* 79, no. 1 (February 2014): 1–16. https://doi.org/10.1177/0003122413515997.

Sage, Daniel, and Patrick Diamond. 'Europe's New Social Reality: The Case Against Universal Basic Income.' *Policy Network*, 2017. http://www.policy-network.net/publications/6190/Europes-New-Social-Reality-the-Case-Against-Universal-Basic-Income.

Santoni de Sio, Filippo, Txai Almeida, and Jeroen van den Hoven. 'The Future of Work: Freedom, Justice and Capital in the Age of Artificial Intelligence.' *Critical Review of International Social and Political Philosophy* (13 December 2021): 1–25. https://doi.org/10.1080/13698230.2021.2008204.

Sen, Amartya. *The Idea of Justice*. Cambridge, MA: Harvard University Press, 2009. https://www.hup.harvard.edu/catalog.php?isbn=9780674060470.

Stilgoe, Jack. 'Machine Learning, Social Learning and the Governance of Self-Driving Cars.' *Social Studies of Science* 48, no. 1 (2017): 25–56. https://doi.org/10.1177/0306312717741687.

Van den Hoven, Jeroen. 'Value Sensitive Design and Responsible Innovation.' In *Responsible Innovation*, edited by Richard Owen, John Bessant, and Maggy Heintz, 75–83. Chichester: John Wiley & Sons, Ltd, 2013. https://doi.org/10.1002/9781118551424.ch4.

Van der Veen, Robert. 'Real Freedom versus Reciprocity: Competing Views on the Justice of Unconditional Basic Income.' *Political Studies* 40, no. 6 (1998): 140–63.

Van Donselaar, G. *The Right to Exploit: Parasitism, Scarcity, Basic Income*. Oxford and New York: Oxford University Press, 2009.

Verbeek, Peter-Paul. *What Things Do: Philosophical Reflections on Technology, Agency, and Design*. University Park, PA: Penn State University Press, 2005.

Walker, Mark. 'BIG and Technological Unemployment: Chicken Little Versus the Economists.' *Journal of Evolution and Technology* 24, no. 1 (2014): 5–25.

Widerquist, Karl. 'Who Exploits Who?' *Political Studies* 54, no. 3 (October 2006): 444–64. https://doi.org/10.1111/J.1467-9248.2006.00614.X.

Willcocks, Leslie. 'Robo-Apocalypse Cancelled? Reframing the Automation and Future of Work Debate.' *Journal of Information Technology* 35, no. 4 (10 December 2020): 286–302. https://doi.org/10.1177/0268396220925830.

Winner, Langdon. *Autonomous Technology. Technics-out-of-Control as a Theme in Political Thought*. Cambridge, MA: The MIT Press, 1977.

———. 'Do Artifacts Have Politics?' *Daedalus* 109, no. 1 (1980): 121–36.

Wright, Erik Olin. 'Basic Income as a Socialist Project.' *Basic Income Studies* 1, no. 1 (4 March 2006). https://doi.org/10.2202/1932-0183.1008.

Zwolinski, Matt. 'A Hayekian Case for Free Markets and a Basic Income.' In *The Future of Work, Technology, and Basic Income*, edited by Michael Cholbi and Michael Weber, 7–26. Routledge, 2019.

3 The design stance in ethics of technology

3.1 The politics of design

Karl Marx once famously wrote: 'The philosophers have only interpreted the world in various ways; the point is to change it.'[1] This must be a constant warning for philosophers of technology. A social and political theory that wants to protect people's freedom should not only analyse current technological development and should instead take what I call a strong, proactive 'design stance' towards it. We must find ways to (a) promote a more inclusive and plural deliberation about the values and principles that should govern new technologies (beyond just efficiency and productivity); this is the idea behind Responsible Innovation;[2] (b) devise specific design frameworks and strategies to define relevant human values to be embedded in technology and to translate them into actionable design requirements; this is the idea of Value-Sensitive Design;[3] (c) support and empowering communities in developing their own existing projects and (technical) expertise and skills; this is the idea of Social and Participatory Design;[4] and (d) devise specific design strategies to spark awareness, reflection, and imagination among designers and the general public, to identify risks, concerns, visions, and opportunities that go beyond the dominant mainstream visions and narratives; this is the idea of Critical and Speculative Design.[5]

Since I will often refer to these four approaches in the rest of the book, the next sections present them in more detail.

3.2 Responsible Innovation

About 13 years ago, the issue of the politics of technology was explicitly addressed in Europe by an intellectual and political movement called Responsible Innovation.[6] The general idea of Responsible Innovation is that technological innovation should be aligned with ethical and social needs as opposed to just following the direction given by technology developers; and that the innovation process should be discussed and negotiated

DOI: 10.4324/9781003303244-4

with the public and so-called 'stakeholders' – people and groups who will be affected by technological development – rather than just being led by policymakers, experts, and *share*holders (people who have financially invested in technological development).

The goal of Responsible Innovation is often presented in terms of broader and more systematic stakeholder participation in critical decisions about emerging and future developments of technology. Stakeholder participation should be organised in ways that allow for more 'inclusivity' and 'reflexivity' of the innovation process, a greater 'responsiveness' of the technological development to societal needs[7] and a stronger 'alignment' of technological development and societal values.[8] The Responsible Innovation movement has successfully influenced the European Commission's innovation policies, and its principles have been explicitly endorsed in many documents and initiatives, such as, for instance, the Horizon 2020 research funding scheme of the EU.

Responsible Innovation proposes a new, more democratic process of innovation. Indeed, one important idea behind the approach, which derives from previous academic and policy approaches such as Technological Assessment, is that stakeholder inclusion can bring two advantages.[9] Firstly, including and engaging stakeholders' views and concerns supports ethics and justice, by giving voice to actors who will be affected by the introduction of technology. Secondly, a broader inclusion of stakeholders should also be pursued for epistemological reasons, as it can contribute to better knowledge, leading to a better capacity to anticipate and manage societal risks. This resonates with the idea of political theory about the epistemological superiority of democracies as the political systems that offer the best terrain for building shared knowledge and for efficiently exchanging information between citizens and policymakers. Experts need the public's knowledge, as it were.[10]

Critics have claimed that promoters of Responsible Innovation have not sufficiently considered the 'politics in deliberation,' that is its concrete social processes and the role of power structures in them. 'How do actors "co-create" outcomes? How do they deliberate? On whose terms is participation (i.e. deliberation) established, and why?'[11] The concern here is that the mere involvement of stakeholders in a deliberation process may not grant all of them the same power to make their values and interests matter and steer the process accordingly.

Also, most promoters of Responsible Innovation are social scientists and policy researchers, with a stronger focus on the policies around technological innovation rather than the design process itself. However, philosopher of technology Jeroen van den Hoven has argued that democratic deliberation over technological development may not be enough and that Responsible Innovation should embed a more explicit design stance to be socially and politically effective. This is because the time when value choices and policy interventions can make a difference is during the actual processes of designing and developing technology. When certain values and stakeholders' interests deeply shape these design and development processes, it is far more difficult, if not impossible, to see to it that technology will serve other values and interests.[12] Design practices and methods are key to making ethical values matter in technological development. Van den Hoven has argued that the two ideas are intrinsically connected and that Responsible Innovation should be realised through Value-Sensitive Design.[13]

3.3 Value-Sensitive Design

In an article from 1996, the US computer scientist Batya Friedman presented the main lines of a theory and an approach that she and her group at the University of Washington started developing in the previous few years: *Value-Sensitive Design*.[14]

According to Value-Sensitive Design, human values should be explicitly and proactively accounted for right from the early stages of and throughout the technology design process. To do so, a multidisciplinary effort is required, and a methodology that includes three elements: a conceptual, an empirical, and a technical analysis. The *conceptual* analysis consists of the theoretical study of the ethical values potentially involved in the technology: what do we want to achieve with this technology (efficiency, wellbeing, justice, etc.) and how do we want to achieve it (through designing for safety, environmental sustainability, etc.)? It also requires an analysis of the stakeholders potentially affected by it: not only 'primary stakeholders' (actors who will directly interact with the technology, typically as users) but also 'secondary' stakeholders (actors who will be affected by the technology indirectly; for example, in the case of an automobile, not only the drivers and passengers but also other road users, the environment, etc.) The *empirical* analysis consists of an empirical study of the stakeholders' values, attitudes, and thinking concerning the specific technology and the specific domain of its application. How can we check what stakeholders think and value about a certain technology? What are their needs, hopes, fears, and attitudes? The *technical* investigation addresses how to translate the specific values and requirements identified in the conceptual and empirical analyses into the technical features of the system.

The Value-Sensitive Design process must be iterative and circular; it does not flow only once and in only one direction. The conceptual investigation influences the empirical and technical research by identifying and defining relevant values, but at the same time abstract concepts are also integrated by stakeholders' interpretations of values in context and by technical consideration about the functioning of the relevant systems. Also, design solutions and prototypes should be conceptually and empirically analysed and assessed, which may lead to new cycles of analyses at all three levels and so on.

Most of the time, the so-called value trade-offs will emerge in the process. Different design options will have a different impact on different values and stakeholders, and it will often be impossible to immediately find design options which will comply with all relevant values of all relevant stakeholders. Often, a point will be reached where it becomes very difficult to increase compliance with one value (of one stakeholder group) without negatively affecting compliance with other values (of other stakeholders). At the same time, many trade-offs and value tensions can be addressed in the design process by applying Value-Sensitive Design.

Friedman illustrated the Value-Sensitive Design approach with the example of an office without windows and the challenge of improving the well-being of the workers who suffer from being deprived of the view of the outside of the building, as confirmed by the empirical studies about workers' attitudes, and scientific studies on workers' well-being.[15] One technical solution would be creating a digital window, a screen displaying real-time images captured by cameras outside. The realisation of the prototype window may confirm the hypothesis but also raise a new ethical (and legal) concern: the privacy of the persons captured by the cameras outside the building. This requires both a check of the legal constraints and of the people's perception of their privacy violation. A new prototype could be made, where cameras would produce low-definition footage that would promote workers' well-being inside the office while also safeguarding the privacy of people outside. But this would not solve the issue of the perception of privacy and the trust of people outside, as they would not know for certain whether and how their image is taken, recorded, and broadcasted. And so, the design process would continue until an ethically satisfactory outcome is reached.

As simple as it may look, Value-Sensitive Design is a revolutionary idea. Many years before, Langdon Winner and others had pointed out that technology is not neutral, and 'artifacts have politics.'[16] The revolutionary aspect of Value-Sensitive Design is that it tries to put this idea into practice and builds theories, methodologies, approaches, and practices to explicitly and systematically (a) analyse the potential values and stakeholders involved in the development of a given technology and (b) propose design solutions to achieve a better – though never perfect and always open to revision – balance of values in technological systems and interactions.

The magnitude of the change can be seen from two perspectives. From the point of view of engineering and design, it amounts to explicitly embedding a process of moral and political deliberation *within* the design process. From the point of view of the ethical, philosophical, and political analysis, it requires this analysis to be framed in design-consequential terms, which can be actionable by technical design.

Therefore, Value-Sensitive Design is a call for philosophers, social scientists, and designers or engineers to rethink their disciplines and open their communities to collaborate in the definition and realisation of social and political values, norms, and principles in and through technological development.

For philosophers, this means getting out from their ivory towers and 'armchair philosophies' and collaborating on translating abstract concepts and values into design principles while remaining open to integrating and revising their theories based on empirical and technical insights. The same applies to anyone interested in normative reflection and political action: lawyers, political activists, and others.

For engineers and designers, it means recognising that their work is intrinsically and irreducibly normative and political, as it consists not only of understanding the world but also, and foremost, of changing it. In *The Sciences of the Artificial*, the philosopher of science and technology Herbert Simon defined design as follows: 'Everyone designs who devises courses of action aimed at changing existing situations into preferred ones.'[17] As there are different preferred future states of affairs, developers should be open to reflect on the possible design directions, on the reasons why they may want to go in one or another direction and what this would mean for the people involved in or affected by this change. They should broaden their concept of design to systematically include an explicit reflection and analysis of the human and societal values involved in the artefacts and systems they contribute to produce.

For social scientists, it means developing new (empirical) methods and disciplines for systematically studying human–technology interactions and the emergence of values in them. Many of these disciplines exist today under the general name of Science and Technology Studies.

Recently, scholars in design have wondered to what extent Value-Sensitive Design has lived up to its ambitions. Notwithstanding their best intentions to realise stakeholder inclusion, designers tend to influence the identification process of relevant values by stakeholders when, for instance, they work with a predefined list of values, which may make them blind to other values held by stakeholders.[18] Also, researchers or designers always maintain the power to present the results of their empirical investigation in their own words. Stakeholders' contributions are often interpreted and paraphrased by the researchers, which, again, even when they act with the best intentions, risks projecting on them the researchers' views and perspectives. Researchers or designers should disclose as much as possible their own values and commitments and how they differ from those of the involved stakeholders. To address this issue, stakeholders should be allowed to 'speak for themselves' as much as possible. In addition to quotations, methods can be used to make stakeholders' voices heard, like allowing or encouraging them to use storytelling, photography, or videos to express their values and ethics.[19]

3.4 Participatory Design

One of the first systematic attempts at realising the ideal of the direct involvement of stakeholders in technological design was called *Participatory Design*. The name originally covered a series of experimental projects in Norway and Sweden from the 1970s.[20] For this reason, this specific historical experience is now sometimes referred to as Scandinavian Participatory Design, to distinguish it from the other, more recent, approaches developed outside the workplace, and discussed below in this section.

In Scandinavian Participatory Design, workers were invited to meetings where they would discuss with managers and engineers the introduction of new technologies in the workplace – typically computers and information systems. The idea was that workers would bring their experience and values to the table and, therefore, influence the development and introduction of the technologies they would eventually be asked to use. In their intentions, this aligns with the ideals of both Responsible Innovation and Value-Sensitive Design. Langdon Winner himself praised Participatory Design in the workplace as an important step towards democratising technology.[21] However, Participatory Design also faced some important issues. Most importantly, it was difficult to ensure that workers had the real power to make their voices and ideas matter in the final managerial choices and even more in the design process. As seen above, similar criticisms have been raised about Responsible Innovation and Value-Sensitive Design. The risk of stakeholders not being able to freely contribute is particularly high in hierarchical contexts like workplaces, where workers may feel the pressure to not challenge the management's decisions and plans. Also, well-intentioned managers and engineers may be unable to effectively communicate the technical aspects of their plans to workers, and these may, in turn, have issues communicating their fears and concerns.

Recently, some design scholars are concerned that any participatory practice that is devised and led by designers, managers, or other experts, even when done with good intentions and not just as a form of 'participation washing,' may still end up creating 'extractivist practices,' where the information collected is mainly used in the interests of the people leading the research project, not the community where the information comes from.[22] In *Design Justice*, Sasha Costanza-Chock insists on the importance of collaborating with communities by recognising and valuing the expertise they already have and supporting it. Rather than seeing stakeholders as a source of precious knowledge and information to be exploited by designers for their own purposes or those of their companies, governments, or others. In this new version of Participatory Design, also called Social Design, the designers offer technical support to realise projects that communities have already started (or tried to start) and where the communities' expertise has already been proven to be available. Present-day Social and Participatory Design has gone beyond the walls of the workplace into communities and civics, with a more explicit social and activist agenda.[23]

3.5 Critical and Speculative Design

Design can also help address another important form of power shaping technological development: the power of dominant cultures, visions, and narratives. To ensure an open and plural technological process, we need to address the domination of certain stakeholders in the design process and certain dominant narratives that make those stakeholders' values accepted in society. By noticing that consumers are often among the fiercest opponents to regulations that would reduce the power of big tech companies, even when this may be in their interest as citizens, US social scientists Culpepper and Thelen wondered: *Are We All Amazon Primed?* [24]

'Critical Design' is a term coined by Anthony Dunne and Fiona Raby in the mid-1990s to describe a design practice that 'uses products to ask questions and raise issues in society and culture.'[25] It typically uses prototype products to prompt questions about possible and (un)desirable futures that technological development can lead to. In a way, it is not far from science fiction and other forms of art where technical production is used to 'push the limits of the lived experience.' It asks not only what kind of technological products we *as stakeholders* want or need, but more broadly, what kind of society we *as people and citizens* want to live in, and what the role of technology should be. In Critical Design, products are designed with the primary intention to push the limits of the lived experience of those interacting with the artefact. The relation between stakeholder participation and artefacts is somehow reversed. Here, the designed object is meant to work for eliciting the stakeholder's understanding and political engagement, rather than the stakeholder's engagement being functional to improving the product.

In line with Langdon Winner's idea of 'epistemological luddism' and the broader idea of critical theory in philosophy and sociology (most famously represented in twentieth-century Western philosophy by the Marx-inspired so-called Frankfurt school), the key idea of Critical Design is that social change should start by becoming fully aware of the forces and structures that shape and steer current society, politics, and economics and of the impact and potential (negative) consequences of accepting these forces and structures. Most importantly, Critical Design is about explicitly recognising that other, more desirable, forms of life are difficult to imagine but nonetheless possible. Critical Design proposes, as it were, a more structural intervention on technology. The idea is that once they have directly experienced certain forms of unfreedom or injustice embedded in current technological development, people may become more willing to engage in various forms of political actions or just change their attitudes, habits, and choices concerning technology. Relatedly, Critical Design practices may influence designers' choices and policymakers' options by contributing to the public debate and deliberation on technological development. In this perspective, Critical Design locates itself somewhere between forms of art or entertainment that explicitly aim at raising public awareness over social and political themes (cf. producing objects with a political goal) and plain tactics of political activism (cf. creating the space for civic engagement and awareness and influencing political choices).

In partial overlap with Critical Design, but possibly with a less-defined political agenda and a more explorative attitude, is the bordering field of Speculative Design, where designers create artefacts, settings, and experiences to prompt people's imagination about alternative presents and speculative futures.[26]

Recently, scholars like Carl DiSalvo and Laura Forlano have tried to connect the traditions of Social and Participatory Design with that of Critical and Speculative. They believe that the two approaches are complementary and can fruitfully complete each other.[27] Critical and Speculative Design have the power to spark reflection and imagination and open new visions, but they have been traditionally practised by professional designers in relative isolation from other social groups and communities and with a relatively low societal impact. Social and Participatory Design have more directly addressed communities and their present issues but possibly without the political, philosophical, and creative breadth of Critical Design. Combining the two may offer ways to create original design experiments that both prompt a reflection on new possible desirable (technological) futures and empower local communities in addressing their issues starting from their visions and expertise.

3.6 Taking a design stance

Design studies and philosophies offer a spectrum of options to realise a design stance in the ethics of technology and promote freedom by design. On the one hand, Value-Sensitive Design offers the possibility to achieve tangible technical responses to ethical and societal challenges raised by technology – systems and products that embed in their functionalities a broader range of values of a broader range of stakeholders. It often does so by keeping designers and experts in the driving seat and involving stakeholders, as it were, as consultants to realise the projects. Non-expert people may not always acquire the power to steer a project in their preferred direction, challenge the project's overall idea, or take a critical stance towards their own needs and values. Also, the product's ethical and political impacts may emerge or become evident only in the use phase, once it is released in the wild. Value-Sensitive Design offers more short-term effectiveness at the possible – though not necessary – cost of less inclusivity and less deep and long-term social change. Participatory Design tries to directly support communities in realising their social goals by developing their technical expertise and supporting their goals. Critical and Speculative Design aims to open up the design process to an even broader participation, with some risk of losing effectiveness and impact in the short run, possibly not being able to contribute to the production of immediately usable objects or systems, but with the long-term goal of creating awareness of the deeper cultural and political problems with technology and the means we may need to address them. This is nicely captured in the idea that 'often times design experiments ... identify and articulate problems that should be addressed, perhaps by people other than designers.'[28]

In the rest of the book, I will suggest that we, as a society, must take a design stance towards technology but do not need to choose once and for all which design philosophy is preferable.[29] We may just need all of them: opening the innovation process to a more direct participation of stakeholders and designing policies to realise a better responsiveness and alignment between them and the innovation processes (Responsible Innovation), promoting design solutions that directly protect and promote human freedom and other human values in the interaction with technology (Value-Sensitive Design) and also design processes that support a freer engagement of people and community, valuing their existing ambitions and expertise (Social and Participatory Design) and a critical reflection that may help people and society liberating from the hidden power of dominant narratives and ideologies or a limited conception of what is possible or desirable (Critical and Speculative Design).

That new ethically or politically driven design practices may promote freedom may sound paradoxical to some ears. Ethical values and democratic procedures are often viewed as a constraint, a limit to the free development of technology. However, the paradox is only apparent. Some ethical and political values are certainly in partial contrast to the dominant model of technological development in industrialised capitalistic economies, and they may, therefore, be perceived as a constraint by those who fully endorse a simple version of this model, one where efficiency and productivity and the will of capital owners and shareholders are the only valid guiding principles of action. However, once they are explicitly accepted as fundamental goals of technological progress, ethical considerations and democratic deliberations may be seen as driving forces rather than limitations; they push designers and developers to break the current models of development and devise *innovative* solutions that can comply with more technical and societal requirements.[30] For instance, a greater awareness of the importance of safety has incentivised the development of new technological directions rather than stopping it, and today many would consider safety as an intrinsic component of good technology, not as a constraint. The same goes for environmental protection. Value pluralism in technology is not a limit to freedom or creativity in technological development, any more than it is in democratic societies more generally. If this sounds untrue, it is because we have been for too long exposed to the cultural propaganda of an all-too-simple capitalistic understanding of technological development.

Notes

1 Karl Marx, *Karl Marx: Selected Writings*, 2nd ed., repr (Oxford: Oxford University Press, 2000), 158.

2 Jack Stilgoe, Richard Owen, and Phil Macnaghten, 'Developing a Framework for Responsible Innovation,' *Research Policy* 42, no. 9 (2013): 1568–80, https://doi.org/10.1016/j.respol.2013.05.008.

3 Batya Friedman, 'Value-Sensitive Design,' *Interactions* 3, no. 6 (Dicembre 1996): 16–23, https://doi.org/10.1145/242485.242493; Jeroen van Den Hoven, Pieter E. Vermaas, and Ibo van de Poel, *Handbook of Ethics, Values, and Technological Design: Sources, Theory, Values and Application Domains* (Dordrecht: Springer, 2015), https://doi.org/10.1007/978-94-007-6970-0.

4 Carl DiSalvo, *Design as Democratic Inquiry: Putting Experimental Civics into Practice* (Cambridge, MA: The MIT Press, 2022); Sasha Costanza-Chock, *Design Justice: Community-Led Practices to Build the Worlds We Need* (Cambridge, MA: The MIT Press, 2020); Peter M. Asaro, 'Transforming Society by Transforming Technology: The Science and Politics of Participatory Design,' *Accounting, Management and Information Technologies* 10, no. 4 (October 2000): 257–90, https://doi.org/10.1016/S0959-8022(00)00004-7.

5 Anthony Dunne and Fiona Raby, *Speculative Everything: Design, Fiction, and Social Dreaming* (Cambridge, MA and London: The MIT Press, 2013).

6 René von Schomberg, *Towards Responsible Research and Innovation in the Information and Communication Technologies and Security Technologies Fields* (LU: Publications Office of the European Union, 2011), https://doi.org/10.2777/58723; Stilgoe, Owen, and Macnaghten, 'Developing a Framework for Responsible Innovation'; Stevienna De Saille, 'Innovating Innovation Policy: The Emergence of "Responsible Research and Innovation",' *Journal of Responsible Innovation* 2, no. 2 (4 May 2015): 152–68, https://doi.org/10.1080/23299460.2015.1045280; Richard Owen, René von Schomberg, and Phil Macnaghten, 'An Unfinished Journey? Reflections on a Decade of Responsible Research and Innovation', *Journal of Responsible Innovation* 8, no. 2 (26 July 2021): 217–33, https://doi.org/10.1080/23299460.2021.1948789.

7 Stilgoe, Owen, and Macnaghten, 'Developing a Framework for Responsible Innovation.'

8 Haico Te Kulve and Arie Rip, 'Constructing Productive Engagement: Pre-Engagement Tools for Emerging Technologies,' *Science and Engineering Ethics* 17, no. 4 (December 2011): 699–714, https://doi.org/10.1007/s11948-011-9304-0.

9 Michiel Van Oudheusden, 'Where Are the Politics in Responsible Innovation? European Governance, Technology Assessments, and Beyond,' *Journal of*

Responsible Innovation 1, no. 1 (2 January 2014): 67–86, https://doi.org/10.1080/23299460.2014.882097.

10 More on Responsible Innovation in Chapter 9 below.

11 Van Oudheusden, 'Where Are the Politics in Responsible Innovation? European Governance, Technology Assessments, and Beyond.'

12 Jeroen van den Hoven, 'Value Sensitive Design and Responsible Innovation,' in *Responsible Innovation*, eds. Richard Owen, John Bessant, and Maggy Heintz (Chichester: John Wiley & Sons, Ltd, 2013), 75–83, https://doi.org/10.1002/9781118551424.ch4.

13 Van den Hoven.

14 Friedman, 'Value-Sensitive Design.' For a presentation of today's state of the art, see van den Hoven, Vermaas, and van de Poel, *Handbook of Ethics, Values, and Technological Design: Sources, Theory, Values and Application Domains*.

15 Batya Friedman et al., 'Value Sensitive Design and Information Systems,' in *Human-Computer Interaction and Management Information Systems: Foundations*, eds. Ping Zhang and Dennis Galletta (New York: Routledge, 2007), 348–372.

16 Langdon Winner, 'Do Artifacts Have Politics?,' *Daedalus* 109, no. 1 (1980): 121–36.

17 Herbert A. Simon, *The Sciences of the Artificial*, 3rd ed. (Cambridge, Massachusetts: MIT Press, 1996), 111.

18 Christopher A. Le Dantec, 'Situated Design: Toward an Understanding of Design through Social Creation and Cultural Cognition,' in *Proceedings of the Seventh ACM Conference on Creativity and Cognition* (C&C '09: Creativity and Cognition 2009, Berkeley California USA: ACM, 2009), 69–78, https://doi.org/10.1145/1640233.1640247.

19 Alan Borning and Michael Muller, 'Next Steps for Value Sensitive Design,' in *Proceedings of the SIGCHI Conference on Human Factors in Computing Systems* (New York, NY: ACM, 2012), 1125–34, https://doi.org/10.1145/2207676.2208560.

20 Asaro, 'Transforming Society by Transforming Technology.'

21 Asaro, 286.

22 Costanza-Chock, *Design Justice*.

23 DiSalvo, 'Design as Democratic Inquiry.'

24 Pepper D. Culpepper and Kathleen Thelen, 'Are We All Amazon Primed? Consumers and the Politics of Platform Power,' *Comparative Political Studies* 53, no. 2 (2020): 288–318, https://doi.org/10.1177/0010414019852687.

25 DiSalvo, 'Design as Democratic Inquiry,' 17.

26 James Auger, 'Speculative Design: Crafting the Speculation,' *Digital Creativity* 24, no. 1 (1 March 2013): 11–35, https://doi.org/10.1080/14626268.2013.767276.

27 Laura Forlano, 'Decentering the Human in the Design of Collaborative Cities,' *Design Issues* 32, no. 3 (1 July 2016): 42–54, https://doi.org/10.1162/DESI_a_00398; DiSalvo, 'Design as Democratic Inquiry.'

28 DiSalvo, 'Design as Democratic Inquiry,' 30.

29 See in particular chapter 10 below.

30 Jeroen van den Hoven, Gert-Jan Lokhorst, and Ibo van de Poel, 'Engineering and the Problem of Moral Overload,' *Science and Engineering Ethics* 18, no. 1 (March 2012): 143–55, https://doi.org/10.1007/s11948-011-9277-z.

References

Asaro, Peter M. 'Transforming Society by Transforming Technology: The Science and Politics of Participatory Design.' *Accounting, Management and Information Technologies* 10, no. 4 (October 2000): 257–90. https://doi.org/10.1016/S0959-8022(00)00004-7.

Auger, James. 'Speculative Design: Crafting the Speculation.' *Digital Creativity* 24, no. 1 (1 March 2013): 11–35. https://doi.org/10.1080/14626268.2013.767276.

Borning, Alan, and Michael Muller. 'Next Steps for Value Sensitive Design.' In *Proceedings of the SIGCHI Conference on Human Factors in Computing Systems*, 1125–34. New York: ACM, 2012. https://doi.org/10.1145/2207676.2208560.

Costanza-Chock, Sasha. *Design Justice: Community-Led Practices to Build the Worlds We Need*. Cambridge, MA: The MIT Press, 2020.

Culpepper, Pepper D., and Kathleen Thelen. 'Are We All Amazon Primed? Consumers and the Politics of Platform Power.' *Comparative Political Studies* 53, no. 2 (2020): 288–318. https://doi.org/10.1177/0010414019852687.

De Saille, Stevienna. 'Innovating Innovation Policy: The Emergence of "Responsible Research and Innovation".' *Journal of Responsible Innovation* 2, no. 2 (4 May 2015): 152–68. https://doi.org/10.1080/23299460.2015.1045280.

DiSalvo, Carl. *Design as Democratic Inquiry: Putting Experimental Civics into Practice*. Cambridge, MA: The MIT Press, 2022.

Dunne, Anthony, and Fiona Raby. *Speculative Everything: Design, Fiction, and Social Dreaming*. Cambridge, MA and London: The MIT Press, 2013.

Forlano, Laura. 'Decentering the Human in the Design of Collaborative Cities.' *Design Issues* 32, no. 3 (1 July 2016): 42–54. https://doi.org/10.1162/DESI_a_00398.

Friedman, Batya. 'Value-Sensitive Design.' *Interactions* 3, no. 6 (December 1996): 16–23. https://doi.org/10.1145/242485.242493.

Friedman, Batya, Peter H. Jr. Kahn, and Alan Borning. 'Value Sensitive Design and Information Systems.' In *Human-Computer Interaction and Management Information Systems: Foundations*, edited by Ping Zhang and Dennis Galletta, 348–372. New York: Routledge, 2007.

Le Dantec, Christopher A. 'Situated Design: Toward an Understanding of Design Through Social Creation and Cultural Cognition.' In *Proceedings of the Seventh ACM Conference on Creativity and Cognition*, 69–78. Berkeley, CA: ACM, 2009. https://doi.org/10.1145/1640233.1640247.

Simon, Herbert A. *The Sciences of the Artificial*, 3rd ed. Cambridge, MA: The MIT Press, 1996.

Stilgoe, Jack, Richard Owen, and Phil Macnaghten. 'Developing a Framework for Responsible Innovation.' *Research Policy* 42, no. 9 (2013): 1568–80. https://doi.org/10.1016/j.respol.2013.05.008.

Te Kulve, Haico, and Arie Rip. 'Constructing Productive Engagement: Pre-Engagement Tools for Emerging Technologies.' *Science and Engineering Ethics* 17, no. 4 (December 2011): 699–714. https://doi.org/10.1007/s11948-011-9304-0.

Van den Hoven, Jeroen, Gert-Jan Lokhorst, and Ibo van de Poel. 'Engineering and the Problem of Moral Overload.' *Science and Engineering Ethics* 18, no. 1 (March 2012): 143–55. https://doi.org/10.1007/s11948-011-9277-z.

Van den Hoven, Jeroen. 'Value Sensitive Design and Responsible Innovation.' In *Responsible Innovation*, edited by Richard Owen, John Bessant, and Maggy Heintz, 75–83. Chichester: John Wiley & Sons, Ltd, 2013. https://doi.org/10.1002/9781118551424.ch4.

Van den Hoven, Jeroen, Pieter E. Vermaas, and Ibo van de Poel. *Handbook of Ethics, Values, and Technological Design: Sources, Theory, Values and Application Domains*. Dordrecht: Springer, 2015. https://doi.org/10.1007/978-94-007-6970-0.

Van Oudheusden, Michiel. 'Where Are the Politics in Responsible Innovation? European Governance, Technology Assessments, and Beyond.' 1, no. 1 (2 January 2014): 67–86. https://doi.org/10.1080/23299460.2014.882097.

Von Schomberg, René. *Towards Responsible Research and Innovation in the Information and Communication Technologies and Security Technologies Fields*. Publications Office of the European Union, 2011. https://doi.org/10.2777/58723.

Winner, Langdon. 'Do Artifacts Have Politics?' *Daedalus* 109, no. 1 (1980): 121–36.

Part 2

Freedom at work

4 Algorithmic domination and workers' freedom

4.1 Workers, robots, and control

'I would rather go back to a correctional facility and work for 18 cents an hour than do that job,' says Jimpat Lacewell, who quit his job as a sorter in the US Amazon distribution centre of Staten Island after only three days because the place reminded him of prison. Lacewell's story is contained in a 2020 reportage from *The Guardian* on the working conditions of Amazon's employees in some US warehouses.[1] Rosa Cummings from Amazon's New York facility explains that sorters must inspect and scan 1,800 packages per hour – 30 per minute. She tells of having been reported for missing her hourly rate, even though she did so to clean up the conveyor belt, after, as sometimes happens, an envelope with liquid had burst onto it. A write-up is a serious affair because after two or three of them you can get fired. Indeed, as Cummings explains, 'People get fired regularly' and 'You can get fired for anything.' Such pressure does not help keeping the place safe and, in fact, the injury rate at the facility was found to be three times the national average for warehouses. Other workers also protested that their medical conditions were not taken seriously and that the 15-minute breaks they were given would not even allow them to walk to and back from the break room. Ilya Geller, a stower, said: 'You're being tracked by a computer the entire time you're there.' And: 'You don't get reported or written up by managers. You get written up by an algorithm … if there is too much time lapsed between items, the computer will know this, will write you up, and you will get fired.' Raymond Velez, employed for about one year at the JFK8 warehouse, protested that their employers 'care more about the robots than they care about employees,' and Juan Espinoza, a picker at the Staten Island warehouse, after quitting due to the gruelling working conditions, commented: 'I am a human being, not a robot.'

It does not have to be that way, though. Technology has enormous potential to also be used for the *good* of workers. Think of exoskeletons or other technological tools that help workers perform heavy tasks like

DOI: 10.4324/9781003303244-6

carrying weights with less pain and risk of injury. Or, think of the use of digital platforms in academic research and much other work during the pandemic. Mass working-at-home has often had a negative impact on the mental well-being of workers and students and exacerbated differences among people by creating issues, especially with persons with more vulnerable psychological and social profiles. At the same time, many workers have had the opportunity to experiment with more favourable work arrangements.[2] In one particular example, the disability studies scholar and campaigner Ashley Shew has claimed that COVID has highlighted the potential of technology to allow for more inclusive workspaces for people with disability and called for more work in that direction.[3]

In *The Poverty of Philosophy* (1847), Marx famously wrote: 'The hand-mill gives you society with the feudal lord; the steam-mill society with the industrial capitalist.'[4] I don't think we should fear that technology may fully determine complex social and political structures, and I'm also not going into the thorny issue of whether this is what Marx thought. But we can safely assume that technological development is a crucial component in shaping the modes of production and power relations at work. As Langdon Winner said: 'Technology has politics.'[5]

Chapter 1 explored the ethical and political concerns of artificial intelligence (henceforth AI) systems replacing workers and leaving them unemployed. This chapter explores the question of how AI can impact the freedom and well-being of employed workers. Recent studies suggest, in contrast to common fears, that AI may in fact not trigger a massive replacement of workers in the next few decades; however, it is very likely to reshape manager–worker and worker–worker relations as well as the way work is done.[6]

How this reshaping may play out is an open question. Based on a review of a diverse body of literature in philosophy and social science, philosopher Jilles Smids and colleagues conclude that there is no simple answer to the question whether the introduction of AI systems at work – in their case: intelligent robots – will make human work better or worse. It just depends on how these systems will look like and how they will be introduced.[7]

To govern the transition to a future where workers' freedom is protected or promoted rather than limited, it is therefore very important to understand how AI in the workplace currently works, how it is changing the relationships of power in the workplace, and what can be done to address oppressive relationships mediated by new technologies.

This chapter addresses these three issues. Section 4.2 presents some relevant findings about the impact of AI on managerial control and workers' freedom in the workplace, as presented in the social science literature. The main result of this section is that AI gives extraordinary new powers to managers to control workers. However, as with past forms of managerial control, workers may still maintain some 'agency,' that is, some capacity for criticising and resisting the control from these systems.[8] Section 4.3 proposes a philosophical analysis of workers' (un)freedom in one particular kind of work, called 'microtasking' or 'gig economy' (e.g. at Uber or Amazon Mechanical Turk). This new AI-mediated control over workers is presented through the lenses of the neo-republican political philosophy, as a form of 'algorithmic domination.'[9] Section 4.4 proposes a design perspective to make sense of and start addressing the issue of workers' free agency in the digital economy. I will present an experiment in Critical Design by Lilly Irani and Six Silberman: Turkopticon.[10] This is a technical tool designed to allow workers to partially resist the domination and exploitation in the gig economy as well as a device to highlight and debunk narratives about digital labour that implicitly support and legitimise domination and exploitation (digital workers as agentless and low-skilled).[11] In Section 4.5, I conclude the chapter with a plea to create a long-term research and policy agenda to better understand and support free agency at work, via multidisciplinary work in philosophy, engineering, social science, and design.

4.2 AI at work: the new contested terrain of control

A multidisciplinary group of researchers from MIT led by Katherine C. Kellogg have recently provided a comprehensive critical review of the literature in social science, assessing the impact of the introduction of AI systems (which they call 'algorithmic technologies') on the structures of managerial control in the workplace.[12] In their review, they discern two types of control: rational control and algorithmic control.

4.2.1 Rational control

Kellogg and colleagues adopt in their review a framework originally proposed by the economist Richard Edwards in his 1979 book *Contested Terrain: The Transformation of the Workplace in the Twentieth Century.*[13] Since the industrial revolution – Edwards had argued – the workplace is the terrain of a constant 'dialectical process' between employers and workers. On the one hand, employers deploy various techniques and technologies to establish 'rational control' over workers, to maximise the value of the workers' labour by making their performance more efficient, increasing productivity and ensuring workers' compliance with the management's interests, plans, and wishes. On the other hand, workers constantly engage in resistance to maintain their autonomy, dignity, and identity.[14] 'Rational control' functions by appealing to the workers' self-interest. Although control can also be obtained by 'winning workers' hearts and minds,' that is by convincing them, Kellogg and colleagues do not discuss this alternative form of control, which they call 'normative.'[15]

Rational control has been historically achieved in two ways. In the first half of the twentieth century, control was exercised mainly via physical and technical means, using machines like in the paradigmatic case of assembly lines, where 'the worker became nearly as much locked in place as the machinery.'[16] Edwards calls this realisation of rational control 'technical control.' Through technical control employers can direct workers without being present on the work floor. Workers are driven by machines to perform certain tasks at a particular rate.[17] When workers in Amazon warehouses protested with the words 'I am not a robot,' they were writing a new chapter in a long tradition of complaints about the excessive mechanisation of human life in the assembly lines, one that was ironically and iconically captured by Charlie Chaplin's 1936 film *Modern Times*.

The second part of the twentieth century witnessed the rise of another realisation of rational control over workers: bureaucratic control. Unlike technical control, bureaucratic control is not achieved by physical means but by rules, protocols, job descriptions, and metrics of evaluation. The distinction between technical and bureaucratic control is an abstract conceptual one. In practice, models of control frequently overlap and can be combined in hybrid forms.[18]

Rational control functions through three related mechanisms: direction, evaluation, and discipline. *Direction* entails the specification of what needs to be performed, in what order and time period, and with what degree of accuracy. *Evaluation* entails the review of workers to correct mistakes, assess performance, and identify those who are not performing adequately. *Discipline* entails the punishment and reward of workers so as to elicit cooperation and enforce compliance with the employer's direction.[19]

4.2.2 Algorithmic control

Moving beyond Edwards' rational control in the twentieth century, and looking at the literature on the use of AI technologies at work, Kellogg and colleagues observe three important facts.

First, the three dimensions of rational control of work identified by Edwards – direction, evaluation, and discipline – are still present in managerial control. However, the use of AI systems has given them a different shape than traditional technical and bureaucratic control. AI has specific features that set it apart from previous technologies.

AI systems are more *comprehensive*, *instantaneous*, *interactive,* and *opaque*. More *comprehensive* since cameras, sensors, and audio devices can now record many data about workers, such as body movement data, biometric data, and sensor data. This allows for new forms of control of workers, including, for instance, screening for drug and alcohol use or collecting emotional and physiological indicators in real time. More *instantaneous* as 'platforms can instantaneously compute, save, and communicate real-time information with workers and managers … As a result, feedback and assessment can be incorporated continuously into the production process.'[20] More *interactive* as employers can, for instance, use algorithmically powered chatbots to monitor chat channels and interactively prompt groups to pause and take a poll regarding the next steps. And *opaquer* for three main reasons: intentional secrecy about the functioning of the systems by the company, difficulty in understanding the systems on the part of workers due to the lack of technical literacy, and intrinsic opacity of machine-learning processes.[21]

If the most iconic representation of the excessive mechanisation of human work in twentieth-century assembly lines was Chaplin's *Modern Times*, the most evocative and dystopian representation of the AI-mediated control in the twenty-first-century workplace is closer to Orwell's Big Brother from his novel *1984*: an invisible master able to acquire real-time, comprehensive, and instantaneous information about workers' bodies and minds, to nudge and prompt them to act through psychological manipulation or to punish and fire them from a distance in case of non-compliance.

Second, by matching Edwards' three mechanisms of rational control – direction, evaluation, and discipline – with the specific functioning of AI technologies, Kellogg and colleagues identify six original forms of workers' control mediated by AI technologies. They call them the '6 Rs.'[22] Employers use AI to direct workers by *recommending* their favourite options via digital interaction and by *restricting* the information that is available to workers in the system; they evaluate workers by *recording* a wide range of fine-grained data about them and their behaviour and by *rating* the workers' performance, including their future performance as predicted via computational technologies; and they discipline workers by *replacing* them, potentially in real time, as soon as their performance is not satisfactory, or *reward* workers interactively and dynamically with more opportunities, higher pay, and others. Table 4.1 summarises the six ways of AI's control over workers found by Kellogg and colleagues in their literature review.

For sure, all of these mechanisms of control existed in some form before AI. However, their algorithmic version has specific features, captured by the above-presented AI novelty: they can be performed by an invisible manager in a more comprehensive, instantaneous, interactive, and opaque

Table 4.1 Kellogg et al.'s six mechanisms of algorithmic control (the 6 Rs)

	Direction	*Evaluation*	*Discipline*
Control mechanisms	Recommending Restricting	Recording Rating	Replacing Rewarding

way. I will give examples of algorithmic restricting, rating, and replacing by the Uber platform in the next section.

A third important result of Kellogg et al.'s study is that, far from being unaware, passive, or powerless subjects of algorithmic control, workers often perceive it as forms of 'coercion' or 'surveillance' or otherwise negatively affecting their freedom or rights. They devise various methods to (partially) resist and counteract these undesirable forms of control. They organise forms of individual resistance via practical actions, such as ignoring the recommendations or gaming the algorithm, or collective actions like forums and platforms to exchange information on strategies to maximise their income in the gig economy. Some have organised platform cooperatives as alternatives to those owned by private companies. Workers have teamed up with activists and computer scientists to raise awareness and fight algorithmic injustices and with lawyers to campaign and lobby for better regulations on privacy, surveillance, and others.[23]

Following the definition of the Edwards,[24] Kellogg and colleagues claim that, after the introduction of AI systems, the workplace remains a 'contested terrain of control.'[25]

In the rest of this chapter, I will elaborate on this algorithmic control as described by Kellogg and colleagues and on how workers can retain the capacity to critically assess the potential oppressiveness and injustice of algorithmic control and to find ways to react and contest it. In the next section I will take a more explicitly normative and philosophical stance and describe some forms of algorithmic control as a form of 'algorithmic domination,' a term coined by neo-republican political philosophy. In Section 4.4, I will take a design perspective and present one project that has tried to support gig workers' resistance to algorithmic domination by design: the Turkopticon.

While the analysis by Kellogg and colleagues applies to a broad range of professional contexts, I will focus mainly on the type of work done in the so-called 'gig economy,' like microtasks done at one's computer (Amazon Mechanical Turk), bike delivery jobs (Deliveroo), and chauffeuring (Uber). The gig economy offers a clear-cut example of systematic attempts to achieve algorithmic control over workers through AI systems. In this case, the companies typically do not provide workers with a physical workspace and the interactions between employers and employees mostly happen on dedicated, AI-powered online platforms. However, much of the proposed analysis may also apply to more traditional forms of works performed in traditional workspaces (shops, factories, offices, hospitals, and others) as well as traditional work performed from home via computers.

Also, while AI can in general be designed and used for good purposes, this chapter mainly focuses on the risks of algorithmic control becoming algorithmic domination. However, as I will explain in the last part of the chapter, a structural change in the way AI is designed and used is possible and desirable.

4.3 Algorithmic domination

Sometimes algorithmic control makes life of workers unfree and miserable. Consider the people employed in the Amazon warehouses in the US as described in the report referred to in the opening of this chapter: feeling imprisoned and surveyed in every movement and decision, subject to the risk of physical and psychological harm, provided with poor medical and psychological support, deprived of any meaningful social interaction, and constantly threatened with losing their job in case of non-compliance. Working in places that seem to combine all the bad parts of the mechanised life of a full-time worker in a twentieth-century's assembly line and of the algorithmically controlled life of the twenty-first-century-AI-mediated work. *Modern Times* reloaded, with a sinister Orwellian twist.

Being exposed to such forms of oppressive control and inhumane working conditions is a gross violation of freedom, according to many – possibly all – modern philosophical and legal approaches. While the philosophical discussion on AI and society can get complicated and controversial, some points are rather straightforward. Any technology that is designed and used in this way is oppressive, these uses should be legally prohibited and sanctioned, and effective policies should be put in place to prevent them.

However, algorithmic control is also realised in contexts that, at least at first sight, do not look so gloomy. In this section I will discuss the so-called gig work like chauffeuring via Uber. These jobs are sometimes presented as allowing for greater freedom for consumers and workers alike. Workers can take jobs and tasks on a case-by-case basis, without having to sign a regular working contract. In the case of Uber, for instance, private car owners can offer rides to customers, fulfilling the function traditionally offered by taxi companies without the need for a license or employment contract. Uber certainly brings some benefits from an economic point of view insofar as they allow maximising the utility of personal assets like cars. They may also increase, at least in the short term, competition and thus the quality and quantity of goods and services available for clients. To some extent, they may have some positive impact on workers, by offering job opportunities to persons who would be otherwise excluded; and on the freedom of work, by giving people a work schedule flexibility they may not enjoy in more traditional jobs. Uber promises workers the possibility to 'work where you want, when you want, and set your own schedule.'[26] As I will explain in the next section, things are more complex than this, and behind the promise of freedom and flexibility, there is also a reality of strict managerial control and domination.

4.3.1 Domination as uncontrolled power supported by 'peripheral agents'

In political philosophy, domination defines a relationship in which an agent or group of agents is in the position to influence or control the behaviour of other individuals or groups, in a way that is unlimited or uncontrolled.[27] While power is the capacity of agent A to influence the behaviour of agent B, domination is a more specific relationship of power, one where A has power over B, and *B cannot control the power of A*. Domination is a subset of power relationships, implying that not every form of social power constitutes domination. Domination is *uncontrolled* power. It is a relationship between an agent that is powerful and one that is vulnerable. An employer has power over the employee – the former being able to tell the latter what to do and how to do it – but this power does not arguably constitute domination if the employee has robust means to resist it, for instance, by being able to negotiate and co-determine their tasks and duties, to push back against unfair requests, or, in a worst-case scenario, to leave their job for a less dominating one, without facing significant economic and existential costs. As discussed in Chapter 1, theories of domination have been made popular in English-speaking philosophy by the so-called neo-republican scholars. The branch of neo-republicanism concerned with freedom, power, and domination in the workplace is called 'labour republicanism.'[28]

Neo-republican thinkers like Dorothea Gädeke and Nicholas Vrousalis have recently argued that domination crucially depends on social norms, institutions, and practices.[29] Social structures not only can be formalised in legal institutions, such as in the case of marriage, slavery, or work contracts, but can also remain embedded in informal norms and practices. Domination is thus not a relation between *two* actors (i.e. dominator and dominated) but between *three* actors: dominator, dominated, and so-called 'regulators.' The latter is described as 'any agent, role, or internalised norm that contributes appropriately to the creation, reproduction, or perpetuation of the constitutive power dyad.'[30] Gädeke calls them 'peripheral agents.'[31] Arguably any form of domination crucially depends on the presence of these silent, less visible agents who support or enable the relationship of domination. Think of oppressive social systems or political regimes, which could not exist without the key support of this third category of agents different than dominators and dominated. Historians and social scientists indeed describe political catastrophes such as the Shoah in terms of triads of perpetrators, victims, and bystanders: people who did not actively take part in the atrocities of, for instance, the extermination of the Jews during World War II but somehow facilitated it by, for instance, not protesting or implicitly supporting the Nazi regime.[32] The theory of structural domination emphasises that, far from being simple bystanders, peripheral agents often do much more than just watching and not intervening: they support with their behaviour the norms, mindsets, languages, and practices that allow the dominators to control their dominated.

I propose to consider socio-technological systems incorporating AI to be 'regulators' or 'peripheral agents' in manager–worker relations, insofar as they embed norms and principles that allow or support dominating (work) relationships.[33] Recall Langdon Winner's maxim about technology in general: 'artifacts have politics.'[34] Add the more recent motto by Lawrence Lessig about the power of digital architectures in shaping human behaviour: 'code is law.'[35] Indeed, even when AI systems play a key role in supporting dominating relationships, like in the case of the gig economy discussed here, they always operate within a broader social, political, and legal system of other supporting norms. 'The use of algorithms in the gig economy modifies an already-existing structural power relationship between workers and bosses.'[36] As the story of Turkopticon presented in the next section shows, narratives and cultural factors are other important supports for dominating relationships.

To be clear, technological rules and practices do not create workers' domination alone; technologies are not autonomous. Strictly speaking, it is the human agents, in this case companies' owners and managers, not the social and technical institutions, who dominate.[37] At the same time, no company owner or manager would ever be able to achieve so much power over so many people without the continuous support of the socio-technical systems of AI.

This is a general truth of human dominating relationships: 'no single agent may ever acquire such robust power over others without it being at least tacitly enabled by others.'[38] Personal power (of the few) is often supported by norms and practices implicitly accepted by the many. That also explains why the dominated persons may not perceive the force of domination as coming directly from the dominators. The workers in the gig economy may feel dominated by the app in the same way as workers in traditional factories may feel dominated by the machines or by the rules of the assembly lines and even more so since today's gig workers hardly know who their bosses are. This is much more than metaphorical talk. Just like women in a patriarchal society and people in illiberal countries may not perceive men or governments as evil dominating agents, as these often do not need to exert their domination directly (e.g. by using or threatening to use force), but they do so through the support of widely accepted social norms and practices. While these structures do not dominate the person alone, they contribute to reproducing the norms and practices that allow the dominators to maintain their power.[39] This is why power often appears faceless, though it is not.

4.3.2 Restricting, rating, and replacing: Uber's algorithmic domination

Consider the case of Uber, one of the most famous actors in what is commonly called the 'platform economy,' if looked at from the point of view of its technical functioning, or the 'gig economy,' if looked at from the perspective of the workers. Companies in the platform or gig economy typically create a digital platform where service providers and customers can quickly and easily meet and exchange services. In the case of Uber, private car owners can offer rides to customers, thereby performing work traditionally done by taxi companies.

Alex Rosenblat is a US ethnographer who has studied Uber through the lens of the drivers' perspectives. She also collaborated with the lawyer Ryan Calo in analysing the social, technical, and legal mechanisms that allow Uber to control and architect the behaviour of drivers and clients. While Uber presents itself as offering a platform where drivers and clients can freely meet and negotiate services, the modes of interaction are crucially constrained and shaped by the structure and functioning of the app. For instance:

> Uber imposes restrictions on the information available to drivers before they accept a trip, which prohibits them from making informed choices about the rides they agree to carry out: For example, Uber has a policy of blind ride acceptance, such that the driver does not know the destination of the passenger (and hence, the remunerative value of the trip) before she accepts it. This practice is touted as a means to ensure system efficiency and prohibit destination-based discrimination.[40]

Recalling Kellogg and colleagues' taxonomy of algorithmic control (Table 4.1), Uber's policy is a clear example of restricting the information available to workers, in order to direct their behaviour. Drivers and clients are often faced with take-or-leave choices in accepting rides and prices. This is a new form of control as, unlike past forms of rational control 'algorithmic control allows the restriction of information to be incorporated instantaneously and covertly into the work process.'[41]

Something similar happens with the presentation and requested acceptance of the general contractual terms written by Uber. To log in to work, drivers must always agree to new terms of service – often without having the time and capacity to review them.[42]

This algorithmic control amounts to algorithmic domination to the extent that drivers have no insight into the content of the choices they are presented with, they cannot influence, change, or contest the general terms of conditions of their employment or the way the system selects and presents the available options for individual rides. Such asymmetrical distribution of power and information grants the company owning the app a strong form of control, not only over the technical functioning of the app but also over the behaviour of its users. Unlike traditional companies, Uber and the like can 'architect virtually every aspect of a digital interaction' and adjust this architecture dynamically based on the constant gathering of information about drivers, clients, traffic conditions, and more.[43] And they do so, without disclosing what the precise features of their control system are. They use information control as a tool to maintain an uncontrolled power over workers.

The company's power is further reinforced by the threat for the drivers of being temporarily or permanently 'deactivated,' an objectifying term that suitably captures the disempowerment of the workers involved:

> If a driver accepts the trip and afterward declines it, her cancellation rating is affected, which can put her job in jeopardy. Drivers who do not meet the threshold requirements that Uber sets for their behaviors, such as a high ride-acceptance rate, a low ride-cancellation rate, and a high passenger rating, risk being 'deactivated' (temporary suspension or permanent firing) by Uber from the platform.[44]

This passage nicely exemplifies two other features of Kellogg and colleagues' taxonomy of algorithmic control (Table 4.1): rate and replace. These practices also qualify as algorithmic domination, because drivers cannot sufficiently understand the criteria for their evaluation, nor can they discuss or contest them.

The impossibility to engage with or contest the company's decision is an intentional feature of the system design. Uber and other companies in the platform economy have tried to avoid presenting themselves as employers, to limit their obligation under labour and tort law.[45] This has been the topic of judicial litigations, and some national tribunals have forced Uber to recognise drivers as employers and to grant them some more rights.[46] However, since only 'a handful of managers oversee thousands of drivers on a global scale,'[47] workers can hardly interact with a human employer and even less get an explanation or possibility to play any role at all in the company's choices.

Applying the domination framework to the digital workplace helps clarify the extent to which algorithmic control may turn into algorithmic domination. This happens when the technological norms and systems governing the platforms support forms of uncontrolled power by employers over vulnerable workers.

4.4 Design for freedom at work

How to address the issue of algorithmic domination in the digital economy? In this chapter I have claimed that algorithmic domination by companies is supported by different institutions and peripheral agents in various ways. First of all, they are supported by current legal and political frameworks. So, a first possible reaction is to look at traditional policy and regulatory tools. These can do much, but not everything. For example, legislators can force Uber to formally comply with their obligations as an employer[48] and to protect its riders' right not to be discriminated against by the algorithms when they strike or take sick leaves.[49] However, companies can arguably always find new technical and legal solutions to elude this control and devise new design features to maintain their power and protect their profits. Therefore, others have proposed to change the platform economy at its roots by moving away from its capitalistic version towards platform cooperativism[50] or platform socialism[51]: creating 'enterprises that are owned and managed by workers, who share the profits of the business, eliminating the role of capitalist bosses as disciplinary agents and exploiters of workers' labor.'[52] In the evocative words of one of the early promoters of this idea, Trebor Scholz: 'cloning the technological heart of Uber' and 'put it to work with a different ownership model, adhering to democratic values.'[53]

In the next section I propose yet a different direction. I will take seriously Edwards' and Kellogg and colleagues' idea of the workplace as a 'contested terrain' where dominated workers maintain the capacity to react to the oppression of the algorithmic control and reclaim their freedom. I will also rely on the importance of a 'design stance' proposed in the previous chapter. These two elements are well-captured in Lilly Irani's and Six Silberman's project called Turkopticon.

4.4.1 Workers' agency and Critical Design: Turkopticon

The Turkopticon project[54] responded to the work practices at Amazon Mechanical Turk (henceforth AMT), a popular platform where companies can request online microtasks to be done for a few dollars per hour. In 2013, Irani and Silberman wanted to provoke workers on AMT to imagine how an improved platform could look like.

Their project was inspired by two main forms of activist design. On the one hand it follows up on the tradition of Scandinavian Participatory Design, where designers partner up with workers to advocate and support them in the negotiation with companies about the computerisation of the labour. The general goal is to achieve a transition to computer-mediated work that would not de-skill workers or otherwise hamper their needs and values.[55] On the other hand, the Turkopticon project squarely sets itself in the tradition of Critical Design, where designers use artefacts and technological systems not only to address practical problems but also to raise awareness about broader social and political issues. One strand of Critical Design – Speculative Design – explicitly aims to support people, in this case workers, in reflecting on the desirability of their interaction with technology and imagining possible futures where technology better serves their and other ethical and political values and interests.[56] All these design methodologies aim to have a broader scope than Value-Sensitive Design, where workers and other stakeholders are often seen as a fundamental source of qualitative empirical data to achieve a more inclusive design of products but less often as agents of the design and active actors of the social and political critiques. However, the first part of the Turkopticon project also included a process that is in practice close to the Value-Sensitive Design approach: the design of a tool that can support specific sets of values, in this case freedom and justice for workers.

As the first step of their project, Irani and Silberman asked workers to compile – through AMT itself – a hypothetical Bill of Rights to reflect on what good work might mean in the gig economy. Recurring themes of complaint that emerged from the survey were: uncertainty about payment, unaccountable and seemingly arbitrary rejections (i.e. non-payment), fraudulent tasks, prohibitive time limits, pay delays, uncommunicative requesters and administrators, costs of employer errors borne by workers, and low pay.[57]

As a second step, in 2009, Irani and Silberman created the Turkopticon, a web application and browser add-on that integrated the AMT interface with reviews from Turkers (as workers on AMT are called). Turkers were already sharing information in their dedicated forums, but Turkopticon integrated this information directly into the AMT platform so that workers would get other Turkers' experiences about companies offering tasks when receiving the offer. While the Amazon interface is designed to allow employers to make the interaction with potential workers as automated as possible and to address workers individually, the Turkopticon software broke this logic by providing workers with the possibility of evaluating and possibly refusing work requests, thanks to real-time access to experiences and recommendations of other Turkers. Design for workers' freedom, as it were.

That concludes the technical fix. As a third step, the critical reflection part, Irani and Silberman tried to understand what the experience of the design and use of the Turkopticon meant for them and the workers, and which broader lessons may be drawn about their understanding of the role of workers and designers in the digital gig-economy. To do so, they decided to start from an external look at their work. They systematically analysed the way the Turkopticon's 'story' was presented in the Western media and to what extent the media representations and narratives were consistent with the workers' and designers' experience and understanding of the project.

In their conversations with Irani and Silberman, Turkers explicitly complained about the recurring trope of all the media articles on AMT and Turkopticon: 'technologists created this problem of the "digital sweatshop," and technologists — the makers of Turkopticon — can come along and fix it.'[58] In this picture, workers are twice agentless: first in relation to the technology that exploits them and second in relation to the technology that rescues them.

Most of the media coverage indeed presented Turkopticon as a *technological* success, in which designers, programmers, and creatives were the only relevant actors. Turkers and other workers who powered these platforms were instead presented as exploited cogs in the plans of the companies, 'toiling in digital sweatshops with little creativity or agency.'[59]

But – as Irani and Silberman explain – this is a very skewed representation of reality. Turkers are often skilled and competent people. Among the ones they met online are laid-off teachers, mobility-impaired professionals, military retirees, agoraphobic writers, undersupported college students, stay-at-home parents, temporarily out-of-work engineers, and Malaysian programmers-in-training. Around 58% of them have a bachelor or higher degree.[60] Employers take advantage of this variety of skills and profiles when they run surveys, commission web articles, virtualise focus groups, get translations, and sort permissible web 2.0 content from policy violations.

Interestingly, Turkers even maintain and repair AMT.

> They help resolve breakdowns through Turkopticon and through their forums. They advise employers about flawed task designs or bugs. They teach each other how to use tools. In short, they do much invisible work, some of which helps AMT as a system appear to work as intended.[61]

Also, Turkers' social and political engagement is big. As Irani and Silberman remark, Turkers operate a variety of forums, employer review sites, and job-sharing platforms. Especially on the two major worker-run forums, they

> share advice with one another, negotiate norms of work, and struggle to establish more interactive and participatory relationships with employers ... operate the forums, fundraise hosting feeds, and moderate communities. They regulate conversations, recruit requesters into collaboration, and educate newcomers. They also might band together to support fellow Turkers in acute need, lending money or prayers.[62]

One of the forums also keeps track of the media coverages of AMT.[63]

Most of the journalists meant well in praising the technical work of the Turkopticon designers and exposing the exploitation of the Turkers.

However, their representation of the Turkopticon designers as smart problem-solvers who came to the rescue of helpless workers doing labour 'programmed by employers with higher-value skills and market-making intuitions'[64] is not only inaccurate but also dangerous. It involuntarily supports the same narrative that makes this exploitation possible and – in the eyes of many – acceptable.

The narrative is one in which 'innovators' – those inventing new tools or running million-user platforms – are the only ones doing important (technological) work, while platform workers are just cogs in the machines, no matter how precious, cognitively demanding and time- and energy-consuming their maintenance and repair contributions are, and irrespective of them doing the invisible 'everyday design' that keeps the platform going and working. In this distorted picture, the fact that companies and designers get all the (economic) rewards while workers are left with crumbs may sound sad but not altogether unfair.

And so, a project that started with engaging with workers in a reflection about the possible improvement of their working conditions took workers and designers to a conclusion that went well beyond the creation of a technological solution. As its inventors found out, Turkopticon was not (only) a solution but rather a symptom of a broader problem.[65] The broader problem was the cultural conditions – the 'stories,' 'narratives,' or 'imaginaries' – that make technological exploitation possible and difficult to address. This awareness also sheds light on the responsibilities of academic designers.

Consistently with this approach, Irani and Silberman's story ends with their own propositions to stop implicitly supporting the dangerous narrative of designer heroes versus helpless workers and to explicitly recognise that workers have more power and agency to tell their story and should be seen as the potential agents of their emancipation. They committed to 'teach journalists a different way to think about people's relationship to technology that let design share the stage with other practices of making, articulation, maintenance, and communication;'[66] they even started telling journalists that they would not contribute to their stories unless they also spoke with workers. In addition, they committed to treat workers as experts rather than just a source of qualitative data and tried to involve them in academic activities. Dynamo, a later project similar in spirit to Turkopticon, follows this lead by more actively including Turkers in the design of the system and inviting Turkers to co-author the paper presenting the project.[67]

4.4.2 *The future of research into the future of work*

The Turkopticon story contains some important lessons for any academic project that wants to proactively and systematically address the issue of the future of work vis-à-vis technological development, with a strong emphasis on workers' agency, freedom, well-being, and other values.

Give power to workers, avoid the 'design saviour' complex. First, it is fundamental to design the participation of the workers into the research programme from the very beginning, avoiding the inclination to design for workers without letting them co-design for themselves. As Irani and Silberman eloquently put it, echoing the infamous complex of the 'white saviour,' designers and academics should avoid the complex of the 'design saviour,' aiming at innovating people out of their problems, rather than supporting them in addressing their problems.[68]

As scholars in Critical Design have emphasised, design can do much but not everything, and often cultural and social protest is more effective than the production of objects.[69] The next chapter will elaborate on this point, by addressing the broader issue of the dominating power of big techs on users, a power that can hardly be resisted by only relying on smart design solutions. So, social and philosophical theory for political change and technical design should go hand in hand in research projects.

Design for social change takes time. Many designers, legal scholars, computer scientists, engineers, social scientists, and philosophers discovered, as Irani and Silberman did, the importance of combining knowledge and practices from philosophy and engineering, and have admirably found original ways to realise such multidisciplinary projects. However, this effort should become more structural in universities and research institutes.

Also, if it is cultural and social change that must be achieved rather than just better or more 'ethical' technological solutions, then a longer time frame should be considered in future research projects. As Irani and Silberman warn in relation to the Turkopticon, it takes time to 'build sufficiently thick relationships with project participants to let critiques of project conveners emerge.'[70] And, as all researchers who have been involved in multidisciplinary research projects well know, it also takes time to build collaborations among scholars from different disciplines, to build a shared language and research agenda, and to produce genuinely shared research.

Further, if academic research wants to have any political and economic impact, it must build up collaboration with businesses that have a genuine interest in experimenting with new forms of the worker–technology relation – a genuine interest as opposed to an interest in some form of 'ethics-washing' to protect their public reputation. Building these collaborations and identifying spaces for experimentation and prototyping also require time, effort, and resources.

Domination is a structural phenomenon that should be contrasted by structural programmes. Finally, even if researchers from different disciplines, workers from different domains, and company managers manage to team up to acquire fundings and create a long-term multidisciplinary research agenda to design for workers' freedom and other relevant ethical and social values, they may still face one last important challenge. Values and their interpretation change over time and across domains, and technologies that serve some values at the time of their design or in one domain may fail to address other emerging values in a different time or domain.[71] So, any ethically or socially driven research project should not only have a long-term horizon but potentially an infinite one.

As discussed in this chapter, big companies have come to dominate workers not only thanks to effective technological solutions but also through building a widespread culture that supports and justifies or at least does not sufficiently oppose their technological and business models. A culture that is widespread among business people and the public, but also among researchers in public-funded research organisations. Recall Culpepper's and Thelen's provoking question: *Are We All Amazon Primed*?[72] I have argued above that domination by the few occurs thanks to the implicit acceptance by the many of the rules and practices supporting domination. This is the case for oppressive political regimes, unjust social structures — and dominating socio-technological visions.

Breaking the domination of big technological companies and their supporting ideologies requires not only new technical solutions and political actions but also the creation of a new type of structural programmes of joint research – as opposed to short-term research projects – of (academic) researchers in engineering, design, social science, and philosophy to engage with workers and companies in creating and supporting new visions of the future of work.

Notes

1 Michael Sainato, '"I'm Not a Robot": Amazon Workers Condemn Unsafe, Grueling Conditions at Warehouse,' *The Guardian*, 2020, https://www.theguardian.com/technology/2020/feb/05/amazon-workers-protest-unsafe-grueling-conditions-warehouse.

2 Jodi Oakman et al., 'A Rapid Review of Mental and Physical Health Effects of Working at Home: How Do We Optimise Health?,' *BMC Public Health* 20, no. 1 (30 November 2020): 1825, https://doi.org/10.1186/s12889-020-09875-z.

3 Ashley Shew, 'Let COVID-19 Expand Awareness of Disability Tech,' *Nature* 581, no. 7806 (7 May 2020): 9–9, https://doi.org/10.1038/d41586-020-01312-w.

4 Karl Marx, *The Poverty of Philosophy*, New ed. (New York: International Publishers, 1992).

5 Langdon Winner, 'Do Artifacts Have Politics?,' *Daedalus* 109, no. 1 (1980): 121–36.

6 Leslie Willcocks, 'Robo-Apocalypse Cancelled? Reframing the Automation and Future of Work Debate,' *Journal of Information Technology* 35, no. 4 (10 December 2020): 297, https://doi.org/10.1177/0268396220925830.

7 Jilles Smids, Sven Nyholm, and Hannah Berkers, 'Robots in the Workplace: A Threat to—or Opportunity for—Meaningful Work?,' *Philosophy & Technology* 33, no. 3 (1 November 2019): 503–22, https://doi.org/10.1007/S13347-019-00377-4.

8 Katherine C. Kellogg, Melissa A. Valentine, and Angéle Christin, 'Algorithms at Work: The New Contested Terrain of Control,' *Academy of Management Annals* 14, no. 1 (January 2020): 366–410, https://doi.org/10.5465/annals.2018.0174.
9 James Muldoon and Paul Raekstad, 'Algorithmic Domination in the Gig Economy,' *European Journal of Political Theory* 22, no. 4 (2022): 587–607, https://doi.org/10.1177/14748851221082078.
10 Lilly C. Irani and M. Six Silberman, 'Turkopticon,' in *Proceedings of the SIGCHI Conference on Human Factors in Computing Systems* (New York, NY: ACM, 2013), 611–20, https://doi.org/10.1145/2470654.2470742.
11 Lilly C. Irani and M. Six Silberman, 'Stories We Tell About Labor: Turkopticon and the Trouble with "Design",' in *Proceedings of the 2016 CHI Conference on Human Factors in Computing Systems* (CHI'16: CHI Conference on Human Factors in Computing Systems, San Jose California: ACM, 2016), 4573–86, https://doi.org/10.1145/2858036.2858592.
12 Kellogg, Valentine, and Christin, 'Algorithms at Work.'
13 Richard Edwards, *Contested Terrain: The Transformation of the Workplace in the Twentieth Century* (New York: Basic Books, 1979).
14 Kellogg, Valentine, and Christin, 'Algorithms at Work,' 369.
15 Kellogg, Valentine, and Christin, 369.
16 Kellogg, Valentine, and Christin, 370.
17 Kellogg, Valentine, and Christin, 370.
18 Kellogg, Valentine, and Christin, 370.
19 Kellogg, Valentine, and Christin, 369.
20 Kellogg, Valentine, and Christin, 371.
21 Kellogg, Valentine, and Christin, 372.
22 Kellogg, Valentine, and Christin, 372.
23 Kellogg, Valentine, and Christin, 391–95.
24 Edwards, *Contested Terrain.*
25 Kellogg, Valentine, and Christin, 'Algorithms at Work.'
26 Alex Rosenblat and Luke Stark, 'Algorithmic Labor and Information Asymmetries: A Case Study of Uber's Drivers,' *International Journal of Communication* 10 (27 July 2016): 3763; Muldoon and Raekstad, 'Algorithmic Domination in the Gig Economy,' 591.
27 Philip Pettit, *Republicanism: A Theory of Freedom and Government* (Oxford: Oxford University Press, 1999).
28 Alex Gourevitch, 'Labor Republicanism and the Transformation of Work ,' *Political Theory* 41, no. 4 (2013): 591–617, https://doi.org/10.1177/0090591713485370.
29 Dorothea Gädeke, 'Does a Mugger Dominate? Episodic Power and the Structural Dimension of Domination,' *Journal of Political Philosophy* 28, no. 2 (1 June 2020): 199–221, https://doi.org/10.1111/JOPP.12202.
30 Nicholas Vrousalis, 'The Capitalist Cage: Structural Domination and Collective Agency in the Market,' *Journal of Applied Philosophy* 38 (Wiley-Blackwell, 2021): 6, https://doi.org/10.1111/japp.12414.
31 Gädeke, 'Does a Mugger Dominate? Episodic Power and the Structural Dimension of Domination.'
32 Raul Hilberg, *Perpetrators, Victims, Bystanders: The Jewish Catastrophe, 1933–1945*, 1st ed. (New York, NY: Aaron Asher Books, 1992).
33 A similar approach is taken by Muldoon and Raekstad, 'Algorithmic Domination in the Gig Economy,' 593–4.
34 Winner, 'Do Artifacts Have Politics?'

35 Lawrence Lessig, *Code and Other Laws of Cyberspace* (New York: Basic Books, 1999).
36 Muldoon and Raekstad, 'Algorithmic Domination in the Gig Economy,' 594.
37 It is a philosophical controversial topic, whether institutions or systems can dominate without any human agent also being in a position to dominate. See Gädeke, 'Does a Mugger Dominate? Episodic Power and the Structural Dimension of Domination,' 212. Without taking a position in this general debate, I only claim that this is not certainly the case for technological systems in the platform economy.
38 Gädeke, 207.
39 Gädeke, 212.
40 Ryan Calo and Alex Rosenblat, 'The Taking Economy: Uber, Information, and Power,' *Columbia Law Review* 117, no. 6 (March 2017): 1661, https://doi.org/10.2139/ssrn.2929643.
41 Kellogg, Valentine, and Christin, 'Algorithms at Work,' 375.
42 Calo and Rosenblat, 'The Taking Economy: Uber, Information, and Power,' 1630.
43 Calo and Rosenblat, 1651.
44 Calo and Rosenblat, 1661.
45 Calo and Rosenblat, 1623–24.
46 Theo Wayt, 'Uber Must Classify Drivers as Employees, Not Contractors: Dutch Court,' *New York Post*, 13 September 2021, https://nypost.com/2021/09/13/uber-must-classify-drivers-as-employees-not-contractors-dutch-court/.
47 Muldoon and Raekstad, 'Algorithmic Domination in the Gig Economy,' 592.
48 Wayt, 'Uber Must Classify Drivers as Employees, Not Contractors.'
49 Natasha Lomas, 'Italian Court Rules against "discriminatory" Deliveroo Rider-Ranking Algorithm,' *TechCrunch* (blog), 4 January 2021, https://techcrunch.com/2021/01/04/italian-court-rules-against-discriminatory-deliveroo-rider-ranking-algorithm/.
50 Trebor Scholz, 'Platform Cooperativism vs. the Sharing Economy,' Medium.com, 2014, https://medium.com/@trebors/platform-cooperativism-vs-the-sharing-economy-2ea737f1b5ad.
51 James Muldoon, *Platform Socialism. How to Reclaim Our Digital Future from Big Tech* (London: Pluto Press, 2022).
52 Muldoon and Raekstad, 'Algorithmic Domination in the Gig Economy,' 600.
53 Scholz, 'Platform Cooperativism vs. the Sharing Economy,' 14.
54 Irani and Silberman, 'Turkopticon.'
55 Peter M. Asaro, 'Transforming Society by Transforming Technology: The Science and Politics of Participatory Design,' *Accounting, Management and Information Technologies* 10, no. 4 (October 2000): 257–90, https://doi.org/10.1016/S0959-8022(00)00004-7.
56 Anthony Dunne and Fiona Raby, *Speculative Everything: Design, Fiction, and Social Dreaming* (Cambridge, MA and London: MIT Press, 2013).
57 Irani and Silberman, 'Stories We Tell About Labor,' 4574.
58 Irani and Silberman, 4577.
59 Irani and Silberman, 4574.
60 Irani and Silberman, 4578.
61 Irani and Silberman, 4581.
62 Irani and Silberman, 4578.
63 Irani and Silberman, 4574.
64 Irani and Silberman, 4578.
65 Irani and Silberman, 4579.
66 Irani and Silberman, 4580.

67 Irani and Silberman, 4580.
68 Irani and Silberman, 4577.
69 Dunne and Raby, *Speculative Everything*.
70 Irani and Silberman, 'Stories We Tell About Labor,' 4582.
71 Ibo van de Poel, 'Design for Value Change,' *Ethics and Information Technology* 23, no. 1 (26 June 2018): 27–31, https://doi.org/10.1007/S10676-018-9461-9.
72 Pepper D. Culpepper and Kathleen Thelen, 'Are We All Amazon Primed? Consumers and the Politics of Platform Power,' *Comparative Political Studies* 53, no. 2 (2020): 288–318, https://doi.org/10.1177/0010414019852687.

References

Asaro, Peter M. 'Transforming Society by Transforming Technology: The Science and Politics of Participatory Design.' *Accounting, Management and Information Technologies* 10, no. 4 (October 2000): 257–90. https://doi.org/10.1016/S0959-8022(00)00004-7.

Calo, Ryan, and Alex Rosenblat. 'The Taking Economy: Uber, Information, and Power.' *Columbia Law Review* 117, no. 6 (March 2017): 1623–90. https://doi.org/10.2139/ssrn.2929643.

Culpepper, Pepper D., and Kathleen Thelen. 'Are We All Amazon Primed? Consumers and the Politics of Platform Power.' *Comparative Political Studies* 53, no. 2 (2020): 288–318. https://doi.org/10.1177/0010414019852687.

Dunne, Anthony, and Fiona Raby. *Speculative Everything: Design, Fiction, and Social Dreaming*. Cambridge, MA and London: The MIT Press, 2013.

Edwards, Richard. *Contested Terrain: The Transformation of the Workplace in the Twentieth Century*. New York: Basic Books, 1979.

Gädeke, Dorothea. 'Does a Mugger Dominate? Episodic Power and the Structural Dimension of Domination.' *Journal of Political Philosophy* 28, no. 2 (1 June 2020): 199–221. https://doi.org/10.1111/JOPP.12202.

Gourevitch, Alex. 'Labor Republicanism and the Transformation of Work.' *Political Theory* 41, no. 4 (2013): 591–617, https://doi.org/10.1177/0090591713485370.

Hilberg, Raul. *Perpetrators, Victims, Bystanders: The Jewish Catastrophe, 1933–1945*, 1st ed. New York: Aaron Asher Books, 1992.

Irani, Lilly C., and M. Six Silberman. 'Stories We Tell About Labor: Turkopticon and the Trouble with "Design".' In *Proceedings of the 2016 CHI Conference on Human Factors in Computing Systems*, 4573–86. San Jose, CA: ACM, 2016. https://doi.org/10.1145/2858036.2858592.

———. 'Turkopticon.' In *Proceedings of the SIGCHI Conference on Human Factors in Computing Systems*, 611–20. New York: ACM, 2013. https://doi.org/10.1145/2470654.2470742.

Kellogg, Katherine C., Melissa A. Valentine, and Angéle Christin. 'Algorithms at Work: The New Contested Terrain of Control.' *Academy of Management Annals* 14, no. 1 (January 2020): 366–410. https://doi.org/10.5465/annals.2018.0174.

Lessig, Lawrence. *Code and Other Laws of Cyberspace*. New York: Basic Books, 1999.

Lomas, Natasha. 'Italian Court Rules Against "Discriminatory" Deliveroo Rider-Ranking Algorithm.' *TechCrunch* (blog), 4 January 2021. https://techcrunch.com/2021/01/04/italian-court-rules-against-discriminatory-deliveroo-rider-ranking-algorithm/.

Marx, Karl. *The Poverty of Philosophy*, New ed. New York: International Publishers, 1992.

Muldoon, James. *Platform Socialism. How to Reclaim Our Digital Future from Big Tech*. London: Pluto Press, 2022.

Muldoon, James, and Paul Raekstad. 'Algorithmic Domination in the Gig Economy.' *European Journal of Political Theory* 22, no. 4 (2022): 587–607. https://doi.org/10.1177/14748851221082078.

Oakman, Jodi, Natasha Kinsman, Rwth Stuckey, Melissa Graham, and Victoria Weale. 'A Rapid Review of Mental and Physical Health Effects of Working at Home: How Do We Optimise Health?' *BMC Public Health* 20, no. 1 (30 November 2020): 1825. https://doi.org/10.1186/s12889-020-09875-z.

Pettit, Philip. *Republicanism: A Theory of Freedom and Government*. Oxford: Oxford University Press, 1999.

Rosenblat, Alex, and Luke Stark. 'Algorithmic Labor and Information Asymmetries: A Case Study of Uber's Drivers.' *International Journal of Communication* 10 (27 July 2016): 27.

Sainato, Michael. '"I'm Not a Robot": Amazon Workers Condemn Unsafe, Grueling Conditions at Warehouse.' *The Guardian*, 2020. https://www.theguardian.com/technology/2020/feb/05/amazon-workers-protest-unsafe-grueling-conditions-warehouse.

Scholz, Trebor. 'Platform Cooperativism vs. the Sharing Economy.' *Medium.com*, 2014. https://medium.com/@trebors/platform-cooperativism-vs-the-sharing-economy-2ea737f1b5ad.

Shew, Ashley. 'Let COVID-19 Expand Awareness of Disability Tech.' *Nature* 581, no. 7806 (7 May 2020): 9. https://doi.org/10.1038/d41586-020-01312-w.

Smids, Jilles, Sven Nyholm, and Hannah Berkers. 'Robots in the Workplace: A Threat to—Or Opportunity for—Meaningful Work?' *Philosophy & Technology* 33, no. 3 (1 November 2019): 503–22. https://doi.org/10.1007/S13347-019-00377-4.

Van de Poel, Ibo. 'Design for Value Change.' *Ethics and Information Technology* 23, no. 1 (26 June 2018): 27–31. https://doi.org/10.1007/S10676-018-9461-9.

Vrousalis, Nicholas. 'The Capitalist Cage: Structural Domination and Collective Agency in the Market.' *Journal of Applied Philosophy* 38 (2021): 40–54. https://doi.org/10.1111/japp.12414.

Wayt, Theo. 'Uber Must Classify Drivers as Employees, Not Contractors: Dutch Court.' *New York Post*, 13 September 2021. https://nypost.com/2021/09/13/uber-must-classify-drivers-as-employees-not-contractors-dutch-court/.

Willcocks, Leslie. 'Robo-Apocalypse Cancelled? Reframing the Automation and Future of Work Debate.' *Journal of Information Technology* 35, no. 4 (10 December 2020): 286–302. https://doi.org/10.1177/0268396220925830.

Winner, Langdon. 'Do Artifacts Have Politics?' *Daedalus* 109, no. 1 (1980): 121–36.

5 Against the exploitation of internet users

5.1 The exploitation of users

The rise of big technological companies making huge profits by offering (often free-of-charge) services on the internet is probably the most prominent economic shift caused by digital technological development in the past 20 years. At the time of writing, in October 2023, eight of the ten richest persons in the world are owners of digital technological companies: Elon Musk (Tesla and Twitter/X), Jeff Bezos (Amazon), Larry Ellison (Oracle), Sergey Brin, Larry Page (Google), Mark Zuckerberg (Meta/Facebook), Bill Gates, and Steve Ballmer (Microsoft).

An old saying that is often repeated to make sense of the economic success of these internet companies goes like this: 'if you are not paying for the product, you *are* the product.' This is usually interpreted as meaning that even if you are not paying with money, you are still paying for the service by giving your personal data to the companies. Data have become the new currency. The saying, I argue in this chapter, captures only half of the story. The other, less emphasised, part is that 'if you're not paying for the product, you are *the worker*.' Data exploitation is only one part of the story of the success of digital platforms as sellers of advertising space, the other crucial part being the exploitation of the vast amount of time spent by people on these platforms to produce these data, in particular, the intense cognitive, emotional, and social work people perform to support the development of the digital technologies and to make online platform successful.

Human freedom, in different senses, is key to understanding this form of exploitation. On the one hand, it is the free mental activities of people – as opposed to mechanical, menial, and pre-programmed work – that create value for digital platforms. On the other hand, users of platforms are more and more exposed to the domination of the companies owning and controlling the platforms, insofar as they depend on them for many of their vital activities: work, social, and emotional relations, gathering information and others, and they are thus vulnerable and exposed to the

DOI: 10.4324/9781003303244-7

uncontrollable and arbitrary will of the platform owners. So, people's work online is at the same time free, in the sense of it being intelligent and voluntary, and unfree in the sense of being performed under domination. This tension, as I will argue, is typical of many relationships of exploitation: relationships that are consensual, beneficial, or even enjoyable for the exploited persons while at the same being grounded in an objectionable relation of domination and an unfair exchange.

Digital technologies have therefore created the possibility of a new form of domination and exploitation. In addition to supporting new forms of structural domination of workers in the so-called gig economy, they have also allowed for new forms of structural exploitation in the digital economy more broadly: the exploitation of users.

This is, in a nutshell, how the exploitation is realised. The main source of profit of big internet companies is the sale to other companies of online advertisement space. These spaces are highly profitable because of the vast amount of time spent by people on the platforms and because of the vast amount of data about users possessed by the platform owners. Third parties can use these spaces to display advertisements that reach millions of people while at the same time being tailored to their individual interests, personality, social positioning, and others as these can be inferred from the users' meticulously recorded online activities.

Shoshana Zuboff, author of *The Age of Surveillance Capitalism*, has documented the moment when Google's managers discovered the potential of this business.

> From the start, Google had collected data on users' search-related behavior as a by-product of query activity. Back then, these data logs were treated as waste, not even safely or methodically stored. ... All the value that users created was reinvested in their experience in the form of improved search ... The year 2001 brought the dot.com bust and mounting investor pressures at Google. Google decided to try and boost ad revenue by applying its already substantial analytical capabilities to the challenge of increasing an ad's relevance to users – and thus its value to advertisers. ... Google would finally repurpose its growing cache of 'useless' behavioral data ... this shift in the use of behavioral data was an historic turning point. ... what I call *behavioral surplus*: data reserves that are more than what is required for product and service improvements ... as a means of generating revenue.[1]

In this chapter I will argue that the new digital economy, as exemplified by digital platforms and services like those of Google and Facebook, allow the few company owners to systematically exploit the work of the many users. The claim is that common users of platforms, and not only contracted workers like those doing microtasks on Amazon Mechanical Turk or deliveries for Deliveroo, are systematically exploited by the companies owning the platforms. This claim is not obvious because, unlike the so-called gig workers, users do not enter a relationship with tech companies out of economic necessity, that is, to make a living. They rather seem to enter this relationship freely and to benefit greatly from it. Also, their activity seems more rewarding and performed more freely. However, the relationship between companies owning digital platforms and their users is closer than it seems to that between these companies and their (exploited) workers discussed in the previous chapter. In other words, in the present-day digital economy, users are dominated and exploited as much as platform workers.

I will make two connected claims. First, the terms of the relationship between users and digital platforms are much less free than it may superficially seem;[2] in fact, in the terms utilised in this book, companies owning digital platforms *dominate* the users whose lives are heavily dependent on their services. Second, companies owning big digital online platforms are exploitative in two ways: they exploit the work of public agencies like research institutes, school, etc. by extracting an unreciprocated gain from the cognitive skills and knowledge created by them; and they exploit the free labour of users of the platforms who produce the traffic and content that make the platforms successful and economically valuable. While the exploitation of distributed cognitive skills and shared knowledge is effectively captured by the literature on 'cognitive capitalism,'[3] the exploitation of the individual users is best captured by a more orthodox Marxist reading of the digital economy,[4] integrated by the so-called 'audience labour' theory.[5] The rest of the chapter elaborates on these two claims in more detail. The final section sketches some directions to regain freedom in a world increasingly dominated by private digital companies. But first, a brief definition of the concept of exploitation and its relation to the idea of domination is needed.

5.2 Domination and exploitation

Exploitation is one of the most controversial and debated topics in moral and political philosophy and economics, at least since the time of Karl Marx. Marx famously claimed that capitalism allows (few) capital owners to systematically exploit the labour of the many and that capitalism is therefore intrinsically wrong. Very roughly, from a Marxian perspective, the gain of the capitalists is exploitative because it is grounded in an unequal exchange of labour. Capital owners use workers' labour to extract economic value from their capital (land, material, and financial resources), but workers are not allowed to benefit from the surplus they contributed to create and only receive what they need to make their living and thus to keep production going. Most of the surplus would go into the pockets of the capital owners.

Some present-day Marxist philosophers claim that not all unequal exchanges of labour are exploitative and that unequal exchange of labour is a necessary but not sufficient condition for exploitation.[6] One group may freely decide to gift the economic products of their collective efforts and labour to another group. Think of the structure of a non-profit organisation that aims, say, to support refugees, where all the economic surplus produced by the organisation, beyond what is necessary to maintain the organisation, is gifted to refugees. Let us further assume that none of the workers in the organisation depend on the organisation to make their living and all workers freely endorse the organisation's policy to give all their income to the cause. Refugees do not exploit the non-profit organisation, even though they benefit from an unreciprocated or unilateral service from them. What makes an unreciprocated or unilateral service objectionable and exploitative is when it is made possible by a structure of domination.[7] In other words, exploitation takes place when those who offer the unilateral service are at the same time subject to the uncontrolled, asymmetrical power of those who gain from this exchange. Typically, in a capitalist system, exploitation happens whenever the capital owners can claim a unilateral service from workers *due to their position of power* vis-à-vis the workers. The workers need the paid job to make their living and the capital owners are the only ones in the position to offer them such jobs. For instance, in developing countries, multinational factory owners can claim minimally paid services in terrible working environments from workers. In richer countries, employers can even claim unpaid labour from workers-in-training. In both cases, workers do not have the means to counterbalance the (economic) power of their employers. They have no capacity to claim better salaries or better working conditions, and no real option to leave these jobs without facing dramatic existential costs, such as not being able to make their living or to support their family, or, in the case of trainees in richer countries, not finding a desirable paid job in the future. Exploitation happens when actors in a position of domination use their power to impose an unequal exchange of labour with actors dominated by them.

The asymmetric or uncontrolled relation of power captured by theories of domination is key to understand the extent to which (capitalist) exploitation is objectionable. Work relationships, even the ones that are intuitively most unfair – think again of the so-called sweatshops producing garments for multinational companies in countries such as Bangladesh – are sometimes consensual and somehow beneficial for the workers. Workers are often not forced to take the jobs and can leave whenever they want. Also, given their poverty, the meagre salary they receive from the companies often leaves them better off than they would be without the job, and these salaries are sometimes higher than the ones they could get in many other jobs in the same country.[8] Similarly, though often in much less dramatic existential circumstances, unpaid trainees in companies and organisations of richer countries often freely choose to take unpaid positions as they may actually learn a lot, enjoy their time as a trainee, and use the experience to start a brilliant professional career. Still, I argue that these freely accepted and mutually beneficial relationships are morally objectionable because they are established within a strongly asymmetrical relationship between powerful and vulnerable actors. Therefore, even when they are consensual and mutually beneficial, these relationships are not the result of exchanges between free agents.

5.3 'Cognitive capitalism': the private exploitation of social cooperation

5.3.1 Life online: gaining freedom, losing freedom

How does the exploitation of the users of online services by digital companies exactly work? A first philosophical interpretation of the economic success of online digital platforms and their exploitative practices is offered by scholars in the cognitive capitalism literature.[9] As I will argue, this interpretation captures many but not all the issues regarding online exploitation.

In a nutshell, cognitive capitalism theorists claim that digital companies are so successful and wealthy because they effectively enclose and monopolise a part of social life and then rent it out to other companies for profit.[10] Their wealth derives from the fact that people gradually transfer their cognitive, communicative, and collaborative activities to digital platforms, and then companies rent out this space rich in attention, social activities, and personal data to other companies for advertising. So, in contrast to one common narrative, Facebook, for instance, does not make a profit by selling people's data to companies, which they rarely do. Facebook rather rents out the digital space where these data are produced. They are a bit like owners of a shopping mall, who do not 'sell customers' to shops but rather rent out spaces in areas with guaranteed high consumer traffic.[11]

So, why would this be a problem for freedom, and where precisely would the exploitation happen? Let's start with freedom. Search engines and social network platforms, for instance, promise to give us a free space to acquire and share information, establish social and emotional connections, acquire new knowledge, etc. To some extent they certainly do. However, in doing so, they also limit our freedom. First, to make the digital business as profitable as possible, platforms must accumulate as much traffic and personal data as possible. Data are profitable in different ways. Companies like Google and Facebook already managed to translate this capital into wealth mainly via advertisements. Other companies like Uber did not yet manage to capitalise on their data but they keep voraciously accumulating it in the hope of becoming a credible and attractive partner for investors, who may come up with new ways to turn data into wealth (for instance, in the case of Uber's geographical and traffic data, the development of self-driving cars).[12]

So, for the digital business to be as profitable as possible, as much traffic and data as possible must be produced. To achieve this, digital interactions must become dominant in our life, potentially displacing non-digital interactions. So, platforms are designed to outcompete non-digital interactions via low- or zero-fee policies and to eventually create a lock-in effect where it is very costly if not literally impossible to leave the platforms and return to non-digital interactions (or to go to other digital spaces). This happens as soon as people become dependent on them for their social, professional, leisure, romantic, intellectual, political, transport, and other activities. As it is sometimes put, for this business model to succeed, digital platforms must try to 'colonize all of social life.'[13]

This is not the same as claiming that digital interactions are necessarily impoverished or always worse than real-life ones, that they make us stupid, etc. This may or may not be the case, depending on the circumstances, the users, etc. In fact, in many circumstances, online activities and interactions are very valuable and profitable for many users. The point here is, again, one of domination. Digital interactions affect human freedom to the extent that they tend to monopolise and colonise our cognitive and social life, dramatically reducing opportunities to pursue these activities outside those platforms and making us vulnerable and dependent on the uncontrolled power of big private multinational companies. One way to capture this dominating relationship is the fact that companies can at any time change the terms and conditions of their services, including asking users to start paying for their services or deleting the users' data stored on their platforms, without people having much power to resist these requests or contest these choices.

Another source of unfreedom comes from the arbitrariness of the power of the private for-profit companies that control the platforms. By being private and quasi-monopolies, they can shape the users' cognitive, emotional, and communicative behaviour in whatever direction is most profitable for their business. As their companies rely on traffic and data, this typically means whatever makes the user (a) spend time/engage with the platform and (b) produce data that can easily be 'legible' to and extractable by the system. The 'Like' button on Facebook and its emotional variations (Love, Anger, etc.) are key examples in this respect. While linguistic conversation is hard to interpret on a massive scale, the Like button has an unambiguous meaning and therefore offers a wealth of usable data to Facebook.[14] Again, this does not necessarily mean that social engagement via social networks is always bad or harmful or otherwise undesirable. Once more, this depends on the circumstances. While users certainly retain some power to co-shape their cognitive, communicative, and social behaviour, companies still retain decisive and uncontrolled power over this interaction.

5.3.2 The exploitation of cognitive capital

To understand exploitation in cognitive capitalism, in particular exploitation by companies owning online digital platforms, we need to take a little step back and look more generally at the economic process of automation of production. Broadly speaking, automation technologies help increase the productivity of economic processes. Industrial machines make production more efficient, computers make services faster, and so on. However, some economists have remarked that a new form of exploitation is present in the process of automation as ongoing in present-day capitalist digital economies.

First, *'automation' does not equal 'autonomy.'* Technologies that allow for automation and the corresponding productivity gain are not created *ex nihilo* and do not work on their own. They are rather the result of the accumulated knowledge and experience of many people: researchers, programmers, designers, and users. Importantly, much of the work behind the production of technology is cognitive, communicative, and social. Think, for instance, of the ways new scientific theories and technological approaches are generated and developed in formal and informal interactions among people, or how new technological products are developed, tested, and introduced in society. This is part of the 'cognitive capital' that makes scientific and technological process possible. AI and robotics are often called 'autonomous technologies' due to their capacity to perform (original, cognitive) tasks without much human supervision, but many scholars – both in social science[15] and robotics[16] – have strongly warned against the use of this terminology. They have indeed exposed the 'myth of autonomous technologies' and called it a dangerous misnomer. Misnomer because it tends to obfuscate the fact that so-called autonomous systems work only thanks to human knowledge, data and social infrastructures like language, rules, and practices;[17] dangerous because it promotes the false image of a technology that 'doesn't need us'[18] and that develops and acts on its own, only relying on its intrinsic capabilities for its success.

Second, in this perspective, AI-based business is the last step of a long process, the transition to 'cognitive capitalism.' In 'Fragment on Machines' of the *Grundrisse*, a text that only recently got the attention of interpreters, Marx claimed that knowledge has become the main productive force in modern capitalism.[19] With the development of new production technologies, capitalism has relied more and more on scientific progress as well as on workers learning, acquiring skills, interacting among them, and exchanging information. Based on a reading of this short text, Italian autonomist Marxists have formulated the concept of 'immaterial labour.'[20] This concept is also central to the idea of 'post-Fordism' or 'cognitive capitalism.'[21] Fordist or industrial capitalism is the type of capitalism before mass media and computers, where the main sources of value were material capital, such as land, raw materials, and machines, on the one hand, and material labour, such as human work in factories and offices, on the other. In a historical shift, we are now moving towards cognitive capitalism, where a key role is played by 'immaterial labour': cognitive and social work that produce things as knowledge, information, communication, relationships, and emotional responses.[22] This shift has started already long ago but has become particularly visible in the last decades with the development of AI- and data-driven digital technologies. A similar point has recently been made by non-Marxist economists such as Ro Naastepad and Jesse Mulder who proposed to move towards what they call an Aristotelian foundation of economics, where 'human free mental activities' rather than just material labour are formally recognised (and rewarded) as a key element in economic production.[23]

So, cognitive capitalism is based on the exploitation of mental and social capital. Technological development and its economic valorisation crucially depend on some human and social pre-conditions: education, language, intelligence, creativity, communicative and social interactions, social and political structures, and others. But these are usually created, supported, and organised by non-profit organisation like (publicly funded) schools, universities, research centres, social, and political institutions as well as by informal institutions like family, communities, social relations, etc. Therefore, companies reaping the benefit of cognitive work are potentially exploiting these other actors and agencies, at least to the extent that they do not recognise and systematically fund these agencies. They take an unreciprocated gain in the interaction with other social agencies. Interestingly, this point is shared by the neo-Marxist Vercellone[24] and the 'Aristotelian' economists Naastepad and Mulder[25]. Both notice the paradox of a production system where companies achieve a significant productivity growth thanks to the cognitive, communicative, and social work done outside their companies but then usually list this productivity growth on the balance sheet of firms as capital – as if this were solely the result of their deployment of intelligent technology and the labour of their employees. In today's economy, most of the companies' gains flow to financial markets rather than being reinvested in the reproduction of free cognitive and social work from which it derives. The cognitive capital produced by public and social agencies is hidden and appropriated under the myth of 'autonomous technology.' In a Marxist dialectic interpretation, this is sometimes interpreted as cognitive capitalism creating the conditions for its own weakening and decline (capitalism draining the source of its wealth). Be that as it may, this process can certainly be described as an unequal, unreciprocated exchange between tech companies at the expenses of the many individuals and groups who have contributed to their wealth with their cognitive and social activities: a form of exploitation.

So, from the point of view of cognitive capitalism theory, big tech companies exploit human cognitive capital in two ways: first, by making an unequal and unreciprocated gain from the knowledge produced outside their companies, which made their technology and their business possible; second, by enclosing, privatising, and commodifying the cognitive and social activities of people in their digital spaces. From this perspective, digital companies create value by rendering a potentially abundant common resource – people's intellectual, emotional, and social activities – scarce by digitally enclosing them in their private space and preventing them to freely develop elsewhere. Then they rent them out this valuable space to other companies.[26]. The dominant and monopolistic aspect – domination – is not an accidental feature of the economic success of the platform. It is precisely the fact that people are left with little or very costly choice to leave the platform and take their lives elsewhere, which makes the platform attractive to investors. They know the platform will never lose clients.[27]

5.4 The limits of cognitive capitalism theory: free labour

5.4.1 If you're not paying for the product, you are the worker

As popular as it has become, especially through the work of Hardt and Negri,[28] the cognitive capitalism interpretation of the digital economy and more specifically the idea of rent as the main source of value creation for tech companies has not fully convinced other (Marxist) scholars. Tiziana Terranova,[29] Christian Fuchs,[30] and others have insisted that what makes digital companies unfairly rich is not only the privatisation and commodification of people's social interactions but also the exploitation of people's 'free labour.'

To be fair, 'free labour' theorists, by and large, agree with some key insights of the cognitive capitalism approach: in the capitalistic internet, people are not free to enter online relations that are not controlled by big companies, and it is this lack of freedom and opportunities for the many users that creates the wealth of the few company owners.

However, Christian Fuchs has one big important point of disagreement with the cognitive capitalism interpretation. In his view, what companies exploit is not only people's free social interactions but also their labour. By pushing people to live more and more of their lives in the companies' digital private spaces, digital companies not only enclose and extract rent from the common good of social interactions but also rather force people to work for them for free to produce more and more of this good.

Free labour theorists point to one important problem with the rent-based theory of profit in the digital space. The digital space, unlike traditional sources of rent like land or real estate, is way more dynamically shaped by its users and requires constant production and maintenance.[31] Owners of the digital space crucially need the users' work to co-create and maintain it. Cognitive capitalists suggested the metaphor of shops in a shopping mall: spaces for rent in an area where significant consumer traffic is guaranteed. But in this case, it seems more as if the clients were not only invited to visit the mall but also forced to contribute to build or maintain it. Every Like on Facebook, every review on Uber, and every feedback on Google Maps is a brick in the construction of the big digital shopping mall.

A crucial component of Fuchs' theory of digital unwaged labour is his application to new media of the theory of 'audience labour,' developed in the 1970s in relation to traditional media like newspapers, radio, and television.[32] According to this theory, audiences crucially support the marketing of products by investing their time, attention, cognitive, social and emotional skills in attending, understanding and interpreting commercial communication.[33] Audiences of commercial shows are at the same time 'transport workers' who work to market things to themselves.[34] Before the internet, audiences would help companies promote their products by reading, listening to, or watching the advertisements, reflecting and making sense of them, possibly circulating and reinforcing their messages in their social circles by talking and discussing about them, etc. But in present-day digital media, people are not only consumers but also active producers of contents – hence the term 'prosumers.' It is their cognitive and communicative work that generates the space for advertisements that grant companies their revenues. So, tech companies in fact commodify prosumers' work and extract profit from their products – the data-rich digital space they contributed to create. Recall the saying, 'if you are not paying for the product, you *are* the product.' This may not be an accurate description of the process. If the product is a crowded and data-rich digital space, then you are rather the unwaged worker who produces this space.

In this sense, nothing is new under the capitalist sun. The digital online space is the new capital, big techs are the new capital owners, and we 'users' are the new exploited digital workers who make the capital valuable for the companies with our work. The important difference is that we do not see ourselves as exploited workers but as free people who benefit from a valuable free service.

There is one last important point in favour of the thesis of users as providers of unwaged labour. Whereas the users' cognitive and social labour is crucial to produce the online space, this production is directly organised by the companies. Cognitive capitalists sometimes explicitly say that companies owning digital platforms have become more like landowners, extracting value from capital without exercising any function in the organisation of labour.[35] But this does not seem to be the case. Facebook, for instance, does not only extract a rent from the digital space they own but also actively organises, shapes, and controls the digital labour process to ensure that they are able to extract as much value as possible from it. Suffice it to think of the huge investments made by Facebook and other social network companies in designing the interfaces to stimulate and optimise the user engagement.

Once again, to be sure, users may also enjoy the use of the platforms and benefit from them, but this is not incompatible with their labour being exploited or unfree. This is a frequent combination in the domain of cultural or social work. People who enjoy their work or have a stronger vocation for it – teachers, scientists, artists, social workers – can easier be made to accept unfair working conditions, and they may often not even perceive themselves as being exploited. This matches Tiziana Terranova's seminal definition of free labour being acts that are '[s]imultaneously voluntarily given and unwaged, enjoyed and exploited.'[36] In fact, users benefiting from Google's search engine, Maps or Translate services do not make their labour any less exploited than that of an unwaged worker forced to contribute to building a road they need to use every day, or someone forced to work as a nurse in a hospital from where they are also receiving a treatment free of charge.

5.4.2 Unmasking exploitation and the materiality of cognitive capitalism

While activity on social network platforms can more easily be mixed up for free, non-exploitative work, other examples make the exploitation more visible. Consider the story of CAPTCHA.[37] As users of Gmail well know, to login into a Gmail mailbox or access other online services users are sometimes asked to show that 'you are not a robot' by performing a relatively simple task that is apparently difficult for artificial agents or so-called bots. It is a security procedure that has revealed effective in distinguishing real human users and potentially malicious artificial agents, digital viruses, etc. In 1998, when CAPTCHA, or 'Completely Automated Public Turing test to tell Computers and Humans Apart,' was invented and patented (albeit under a different name), the required task was to read a visually distorted string of random alphanumeric characters or simple math questions, and then type in that string or the answer to the questions. However, in 2008 a group of computer scientists introduced an important variation to the program, which eventually became reCAPTCHA. They substituted the random alphanumeric characters, distorted images, and simple math questions with poorly legible characters from books and newspapers, which the user was asked to read and type in. This was a key moment, as they added a second goal to the original CAPTCHA concept. Users not only contribute to the security of the system but also contribute to a digitisation project: the transfer of paper books and magazines into digital form. Without knowing, users were contributing to the development of software for Optical Character Recognition (OCR) in a crowdsourced digitisation project.

In 2009 Google bought the reCAPTCHA program. Since then, they have used it for its own purpose of digitising paper media and creating searchable archives as well as to improve Google Maps' Street View – with users being asked to read actual house numbers from pictures that Google need to complete their maps – and to train machine learning systems. Google LLC now also owns the original patent for the software. Google's reCAPTCHA is currently used by more than 4 million websites. Of the top 10,000 most visited sites, 46.5% use this software.[38] Whenever they need to access their Gmail mailbox, users are literally forced to do a microtask for Google. This is in fact microwork like the one done by Turkers for little money at Amazon Mechanical Turk. Only, it is forced onto internet users and done by them for free. If clicking of the Like button on Facebook might be more easily mistaken for a completely free and voluntary engagement, Google's reCAPTCHA makes labour exploitation of users more clearly recognisable.

Like with the online work of the Turkers, the work of internet users seems to raise a paradox as well. Much of our online work is free, in the sense that it is *not* mechanic, menial, standardised, repetitive, predetermined, or directly controlled by companies but instead cognitive, communicative, and social work that requires knowledge and various complex skills. But it is also unfree as it is done under domination and exploitation. In fact, users lack 'the freedom to enter online relations that are not controlled by companies,' 'ownership and control of corporate online platforms,' and 'control over expressing their experiences online independently from the control of companies.'[39] This multiform lack of freedom, ownership, and power by the users is in turn the source of power for the company, which gives them the possibility to extract an unfair gain from their users through their free labour.

Finally, acknowledging the labour dimension of the digital economy and cognitive activities is also important as part of a broader philosophical task. We should not only highlight the novelty of the digital, post-Fordist, and cognitive-capitalistic economy but also recognise how it continues older forms of capitalistic economies. Knowledge and information are in themselves a special kind of good: unlike land or food, they are 'non-rival' goods that can be enjoyed simultaneously by a potentially unlimited number of people without being consumed. However, the *production, distribution,* and *consumption* of knowledge are subject to material constraints and are therefore not unlimited. From the point of view of companies, software runs on hardware that occupies huge physical spaces and consumes massive amounts of energy, costing money, and impacting the environment; software programmers and developers have limited time and energy so that only big tech can afford to pay them, giving big tech a strong competitive advantage over alternative platforms.[40] As for users, their cognitive, alleged 'immaterial' labour is just as much physical and material, as we are not disembodied minds, our minds work within the physical limitations of our bodies and brains, and our cognitive work impacts our physical and psychological well-being. Our time is limited, and the time we spend doing things online is time we cannot spend in other ways. Calling this work productive labour helps to see this as part of a big social and economic structure that relies not only on the free circulation of knowledge and information and on social cooperation but also on the extraction of raw materials and physical work, the consumption of materials and energy, the labour of hired programmers, and, indeed, the unwaged and often unaware labour of the users.

5.5 Five ideas to regain freedom

How can we regain freedom? In what follows, I will sketch some potential options, yet to be assessed and developed in more detail. These are presented, as it were, in order of proximity to the existing world, from the closest to the farthest – one may be tempted to say, from the least to the most radical. However, as it happens, what may sound like the most radical proposal may turn out to be, all things considered, the most appropriate as well as the most feasible one.

5.5.1 A technical fix

One relatively easy technical solution to fight the massive unfair enrichment of big tech companies would be that of limiting *by design* the possibility to make profit with (targeted) advertising. There already exist possibilities to block ads on browsers through specific software. But now these should be voluntarily installed and activated by users. If this option were made the default on browsers and computers, a big part of the companies' revenues would be lost.[41] However, this would only reduce one current source of economic gain for companies. But recall that exploitation is made possible by domination. Like the case of Uber shows, companies who manage to dominate a certain domain of life may be able to find other ways to attract investments even when they are not making any actual gain;[42] and, as illustrated by the CAPCHA story, dominating companies like Google may also find ways other than advertisements to economically exploit the vulnerability of users.

5.5.2 A union of users

In line with a Marxist analysis, if users are the new exploited workers, they need to become aware of this predicament, organise themselves, and fight against their exploitation.[43] They should stop seeing themselves as partners in a free exchange of services and become aware of their role in generating the wealth of companies. If, for instance, they were to unite in some sort of union, or maybe join existing unions or associations, they may be able to organise a strike; in this way, they would immediately realise that the companies need their work as much as they need the companies' services, and they may gain the power to influence the companies' policies and business. Having regained that power, what would they ask for?

A first request could be that of *economically rewarding the users' labour*. In fact, this request has already been made: a class-action lawsuit of Facebook users against the company's use of content published by them on their pages as to support their products in ads, which appear on their friends' News Feeds. The users involved in the lawsuit explicitly attempted to redefine their participation in social media in terms of work: they demanded more control over the information they generate and ownership over a greater share of the surplus value they produce.[44] Similarly, in January 2015, a US citizen filed a lawsuit against Google for not compensating the work done in filling in the reCAPTCHA.[45] Both lawsuits were dismissed. But even if such complaints were successful, they may have delivered economically underwhelming results. Some years ago, cognitive capitalism scholars calculated the actual rate of exploitation for the providers of free labour on Facebook. They did so to criticise the 'free labour' thesis and to support their rent-based explanation of the wealth of digital companies. They divided Facebook's advertising revenues in 2010 (355 million dollars) by the number of Facebook users at the time (around 500 million), resulting in an average contribution of 0.7 dollars per year per user.[46] It is not clear to what extent this calculation is sound or should count as a knockout argument against the validity of the 'free labour' thesis, as was the intention of the authors who proposed it. To mention only one issue: the CAPTCHA example shows that advertising revenues are not the only way for tech companies to capitalise on their users' labour. What the calculation may suggest is that a battle for the recognition of the individual economic contribution of the users would partially miss the point. Once more, the possibility of economic exploitation is a symptom of a bigger political problem: the domination of the big tech companies. The fact that they may have not yet found a way to better exploit the work of users, or, given the huge number of users they can exploit, that they may remain happy with a minimal individual exploitation, does not make their power smaller or the unfairness of their aggregate massive gain less objectionable.

5.5.3 Platform cooperativism

The lawyer Jeanelle Orsi, who has been imagining ways to create a sharing economy not dependent on big technological companies for many years, has proposed a new model for a house-sharing platform. Trebor Scholz, a strong supporter of the idea of replacing 'platform capitalism' with 'platform cooperativism,' stated that Orsi's proposal is in line with his idea of 'cloning the technological heart of Uber' and 'put it to work with a different ownership model, adhering to democratic values.'[47] Here is Orsi's proposal: 'Imagine that 20 or more large cities collaborate to develop software ("Munibnb") with all the functionality of Airbnb, and then mandate that all short-term rentals be arranged through the municipal platform.'[48] Cities may mandate that short-term rentals must go via this portal, to maintain better control of the transactions and to gain funds. As in the Uber model, customers become suppliers, as the company owning and running the platform does not possess any physical infrastructure. Also, the ecosystem could allow other actors to benefit, for instance, the population who is not involved in the short-rental business, provided that cities invest some of the platform gains in initiatives to engage and support these other actors. However, the control structure could be completely different than in AirBnB. A city could decide to manage housing supply in the interest of a broad range of citizens, for instance by enforcing a cap on rental nights or setting fees that fluctuate with house scarcity.[49] The software could be optimised to respond to the intentions of the city planners and administrators, which in turn could reflect the interest of more citizens than just the hosts. At the same time the hosts could get a higher share of the rental fee than on AirBnB due to the lower transaction costs. Also, part of the gains could be reinvested in projects that interest and benefit other stakeholders, i.e. citizens who do not own any rental property. Finally, and very importantly, the city could remain accountable for the functioning of the system and subject to a duty to open its functioning to citizens, thereby avoiding accountability gaps typical of private internet platforms.

One open question is to what extent citizens would be in practice able to understand the functioning of the platform and to use it to be empowered in their interest as well as to hold the organisation accountable for their (investment) choices and to steer these according to their interest. In any case, experiments of platform cooperativism have so far not been particularly successful, at least compared to their capitalist counterparts, and this only shows what we already know: that big techs have already gained such a dominant position that it is very difficult to get users away from them. Also, their business model is so lucrative that it is difficult to compete with them from a financial point of view – they will often be able to rely on more and better resources, in terms of material equipment, programmers' labour, investments, etc.[50] At the same time, this is not a unique or completely new challenge. Think of the challenge of supporting business models that preserve the environment or respect workers' rights and well-being in a global context where the biggest and most powerful companies do not work in this way. Far from being a reason to give up, the urgency and seriousness of these challenges should motivate us to try harder.

5.5.4 Universal basic freedom

In the end, it may then well be that the most revolutionary proposal to address the domination and exploitation in technological capitalism is also the most feasible, simplest, and oldest one: dramatically increasing the funding and the support of public, non-profit educational, intellectual, scientific, social, leisure activities, etc., possibly by taxing the profits of the big techs. This chapter has highlighted two kinds of exploitation by big tech companies: the unreciprocated exploitation of the public resources that contributed to create the knowledge and skills that support digital technologies and their success and the exploitation of the cognitive work of the users of the platforms. I have claimed that while the economic exploitation is certainly a problem in terms of justice, the main threat to freedom is the domination that makes exploitation possible and is in turn reinforced by it. Technical and political action should address this domination.

The challenge has two sides: how to empower individuals to pursue a cognitive and social life outside the for-profit digital space and how to empower a fairer balance of power between public and private agencies in producing knowledge and communication. Recall the philosophical debate over Universal Basic Income in Chapter 2. Supporters of Universal Basic Income see it – among other things – as a tool to empower (unemployed) people to invest in their own goals, capacities, talents, and interests. I expressed my scepticism that a modest financial sum would be sufficient to support the cognitive and social skill developments, self-esteem, and social connections necessary to achieve those goals. Now I can expand this argument to cover not only people left unemployed by automation but all people potentially exposed to the domination and life colonisation of digital platforms, that is, a significant part of the world population. A little financial support to individuals would not make private companies any less powerful and individual people any less vulnerable in shaping their lives and the investment of their time and efforts. People need also socio-technical structures that will enable them to pursue alternative forms of cognitive and social engagement than those offered by privately owned big digital platforms. This would give them a real choice to avoid the domination of the digital platforms.

This brings us to the other side of the challenge. The capital freed by technology-, data-, and unwaged work-driven productivity growth should be rather reinvested in fuelling the source of free knowledge and socialisation, that is public schools, universities, and other spaces where intellectual and communicative activities can be devised and pursued outside the direct control of private interests. As it happens in healthy democratic societies, power should fund and support people's capacity to contest it. As it were, big private companies must fund the breaking of their own domination. Rather than just investing in a Universal Basic Income, we must also invest in structures to promote *universal basic freedom*.

Notes

1 Shoshana Zuboff, 'Surveillance Capitalism and the Challenge of Collective Action,' *New Labor Forum* 28, no. 1 (24 January 2019): 12–13, https://doi.org/10.1177/1095796018819461.
2 Tim Christiaens, *Digital Working Lives: Worker Autonomy and the Gig Economy* (Lanham: Rowman & Littlefield, 2023); Christian Fuchs, 'The Digital Labour Theory of Value and Karl Marx in the Age of Facebook, YouTube, Twitter, and Weibo,' in *Reconsidering Value and Labour in the Digital Age*, eds. Eran Fisher and Christian Fuchs (London: Palgrave Macmillan UK, 2015), 26–41, https://doi.org/10.1057/9781137478573_2.
3 Christiaens, *Digital Working Lives*.
4 Fuchs, 'The Digital Labour Theory of Value and Karl Marx in the Age of Facebook, YouTube, Twitter, and Weibo.'
5 Dallas Walker Smythe, 'Communications: Blindspot of Western Marxism,' *Canadian Journal of Political and Social Theory/Revue Canadienne de Théorie Politique et Sociale* 1, no. 3 (1977), https://journals.uvic.ca/index.php/ctheory/article/view/13715.
6 Nicholas Vrousalis, 'How Exploiters Dominate,' *Review of Social Economy* 79, no. 1 (2019): 103–30, https://doi.org/10.1080/00346764.2019.1618483.
7 Vrousalis.
8 Benjamin Powell, *Out of Poverty: Sweatshops in the Global Economy*, Cambridge Studies in Economics, Choice, and Society (New York, NY: Cambridge University Press, 2014).
9 Yann Moulier Boutang, *Cognitive Capitalism* (Cambridge: Polity Press, 2011).
10 Christiaens, *Digital Working Lives*, 49.
11 Christiaens, 43.
12 Christiaens.
13 Christiaens, 47.
14 Christiaens, 46.
15 Jack Stilgoe, 'Machine Learning, Social Learning and the Governance of Self-Driving Cars,' *Social Studies of Science* 48, no. 1 (2017): 25–56, https://doi.org/10.1177/0306312717741687.
16 J. M. Bradshaw et al., 'The Seven Deadly Myths of "Autonomous Systems,"' *IEEE Intelligent Systems* 28, no. 3 (May 2013): 54–61, https://doi.org/10.1109/MIS.2013.70; David A. Mindell, *Our Robots, Ourselves: Robotics and the Myths of Autonomy* (New York: Penguin, 2015).
17 Stilgoe, 'Machine Learning, Social Learning and the Governance of Self-Driving Cars.'
18 Bill Joy, 'Why the Future Doesn't Need Us,' *Wired*, January 2000, https://www.wired.com/2000/04/joy-2/.
19 Karl Marx, *Grundrisse: Foundations of the Critique of Political Economy*, Penguin Classics (London: Penguin Books, 1993).
20 Maurizio Lazzarato, 'Immaterial Labour,' in *Radical Thought in Italy* (Minneapolis, MN: University of Minnesota Press, 1996), 133; Christian Fuchs and Sebastian Sevignani, 'What Is Digital Labour? What Is Digital Work? What's Their Difference? And Why Do These Questions Matter for Understanding Social Media?,' *TripleC* 11, no. 2 (6 June 2013): 256, https://doi.org/10.31269/triplec.v11i2.461.
21 Carlo Vercellone, 'From Formal Subsumption to General Intellect: Elements for a Marxist Reading of the Thesis of Cognitive Capitalism,' *Historical Materialism* 15, no. 1 (2007): 13–36, https://doi.org/10.1163/156920607X171681; Moulier Boutang, *Cognitive Capitalism*.

22 Michael Hardt and Antonio Negri, *Multitude: War and Democracy in the Age of Empire* (New York, NY: Penguin Books, 2005), 113.
23 C. W. M. Naastepad and Jesse M. Mulder, 'Robots and Us: Towards an Economics of the "Good Life,"' *Review of Social Economy* 6764 (2018): 1–33, https://doi.org/10.1080/00346764.2018.1432884.
24 Carlo Vercellone, 'From the Crisis to the "Welfare of the Common" as a New Mode of Production,' *Theory, Culture & Society* 32, no. 7–8 (1 December 2015): 85–99, https://doi.org/10.1177/0263276415597770.
25 Naastepad and Mulder, 'Robots and Us: Towards an Economics of the "Good Life."'
26 Christiaens, *Digital Working Lives*, 45.
27 One related concern not discussed here is that even powerful actors like States, as soon as they become dependent on the services of big tech companies, may lose the power to claim their due or to protect the rights and interests of people. See the recent case of Google threatening to withdraw its search engine from Australia to pressure them to withdraw a regulation that would force the company to share royalties with news publishers. 'Google Threatens to Withdraw Search Engine from Australia,' *BBC News*, 22 January 2021, sec. Australia, https://www.bbc.com/news/world-australia-55760673.
28 Hardt and Negri, *Multitude*.
29 Tiziana Terranova, 'Free Labor,' *Social Text* 18, no. 2 (2000): 33–58, https://doi.org/10.1215/01642472-18-2_63-33.
30 Fuchs and Sevignani, 'What Is Digital Labour? What Is Digital Work? What's Their Difference? And Why Do These Questions Matter for Understanding Social Media?'
31 Fuchs, 'The Digital Labour Theory of Value and Karl Marx in the Age of Facebook, YouTube, Twitter, and Weibo,' 33.
32 Smythe, 'Communications: Blindspot of Western Marxism.'
33 A more recent, related concept is 'attention economy,' see e.g. Michael H. Goldhaber, 'The Attention Economy and the Net,' *First Monday*, 7 April 1997, https://doi.org/10.5210/fm.v2i4.519.
34 Fuchs, 'The Digital Labour Theory of Value and Karl Marx in the Age of Facebook, YouTube, Twitter, and Weibo,' 30.
35 Carlo Vercellone, 'The Crisis of the Law of Value and the Becoming-Rent of Profit: Notes on the Systemic Crisis of Cognitive Capitalism,' in *Crisis in the Global Economy: Financial Markets, Social Struggles, and New Political Scenarios*, eds. Andrea Fumagalli and Sandro Mezzadra (Cambridge, MA: The MIT Press, 2010), 100, https://api.semanticscholar.org/CorpusID:155386615.
36 Terranova, 'Free Labor,' 33.
37 Vino Avanesi and Jan Teurlings, '"I'm Not a Robot," or Am I?: Micro-Labor and the Immanent Subsumption of the Social in the Human Computation of ReCAPTCHAs,' *International Journal of Communication* 16 (27 February 2022): 19.
38 Avanesi and Teurlings, 1450.
39 Fuchs and Sevignani, 'What Is Digital Labour? What Is Digital Work? What's Their Difference? And Why Do These Questions Matter for Understanding Social Media?,' 262.
40 Fuchs and Sevignani, 271–72.
41 Fuchs, 'The Digital Labour Theory of Value and Karl Marx in the Age of Facebook, YouTube, Twitter, and Weibo.'
42 Christiaens, *Digital Working Lives*.

43 Fuchs, 'The Digital Labour Theory of Value and Karl Marx in the Age of Facebook, YouTube, Twitter, and Weibo.'
44 Eran Fisher, 'Class Struggles in the Digital Frontier: Audience Labour Theory and Social Media Users,' *Information, Communication & Society* 18, no. 9 (2 September 2015): 1108–22, https://doi.org/10.1080/1369118X.2015.1018300.
45 Avanesi and Teurlings, '"I'm Not a Robot," or Am I?: Micro-Labor and the Immanent Subsumption of the Social in the Human Computation of ReCAPTCHAs,' 1441.
46 Adam Arvidsson and Elanor Colleoni, 'Value in Informational Capitalism and on the Internet,' *The Information Society* 28, no. 3 (1 May 2012): 138, https://doi.org/10.1080/01972243.2012.669449; Christiaens, *Digital Working Lives*, 42–43.
47 Trebor Scholz, 'Platform Cooperativism vs. the Sharing Economy,' Medium.com, 2014, 14, https://medium.com/@trebors/platform-cooperativism-vs-the-sharing-economy-2ea737f1b5ad.
48 Nathan Schneider and Jeanelle Orsi, '5 Ways to Take Back Tech,' *TheNation.com*, 2015, https://www.thenation.com/article/archive/5-ways-take-back-tech/.
49 Schneider and Orsi.
50 Fuchs and Sevignani, 'What Is Digital Labour? What Is Digital Work? What's Their Difference? And Why Do These Questions Matter for Understanding Social Media?,' 271–72.

References

Arvidsson, Adam, and Elanor Colleoni. 'Value in Informational Capitalism and on the Internet.' *The Information Society* 28, no. 3 (1 May 2012): 135–50. https://doi.org/10.1080/01972243.2012.669449.

Avanesi, Vino, and Jan Teurlings. '"I'm Not a Robot," or Am I?: Micro-Labor and the Immanent Subsumption of the Social in the Human Computation of ReCAPTCHAs.' *International Journal of Communication* 16 (27 February 2022): 19.

BBC News. 'Google Threatens to Withdraw Search Engine from Australia.' 22 January 2021, sec. Australia. https://www.bbc.com/news/world-australia-55760673.

Bradshaw, J. M., R. R. Hoffman, M. Johnson, and D. D. Woods. 'The Seven Deadly Myths of "Autonomous Systems".' *IEEE Intelligent Systems* 28, no. 3 (May 2013): 54–61. https://doi.org/10.1109/MIS.2013.70.

Christiaens, Tim. *Digital Working Lives: Worker Autonomy and the Gig Economy. Off the Fence: Morality, Politics and Society*. Lanham: Rowman & Littlefield, 2023.

Fisher, Eran. 'Class Struggles in the Digital Frontier: Audience Labour Theory and Social Media Users.' *Information, Communication & Society* 18, no. 9 (2 September 2015): 1108–22. https://doi.org/10.1080/1369118X.2015.1018300.

Fuchs, Christian. 'The Digital Labour Theory of Value and Karl Marx in the Age of Facebook, YouTube, Twitter, and Weibo.' In *Reconsidering Value and Labour in the Digital Age*, edited by Eran Fisher and Christian Fuchs, 26–41. London: Palgrave Macmillan, 2015. https://doi.org/10.1057/9781137478573_2.

Fuchs, Christian, and Sebastian Sevignani. 'What Is Digital Labour? What Is Digital Work? What's Their Difference? And Why Do These Questions Matter for Understanding Social Media?' *TripleC* 11, no. 2 (6 June 2013): 237–93. https://doi.org/10.31269/triplec.v11i2.461.

Goldhaber, Michael H. 'The Attention Economy and the Net.' *First Monday*, 7 April 1997. https://doi.org/10.5210/fm.v2i4.519.

Hardt, Michael, and Antonio Negri. *Multitude: War and Democracy in the Age of Empire*. New York: Penguin Books, 2005.

Joy, Bill. 'Why the Future Doesn't Need Us.' *Wired*, January 2000. https://www.wired.com/2000/04/joy-2/.

Lazzarato, Maurizio. 'Immaterial Labour.' In *Radical Thought in Italy*, edited by Paolo Virno and Michael Hardt, 133–46. Minneapolis, MN: University of Minnesota Press, 1996.

Marx, Karl. *Grundrisse: Foundations of the Critique of Political Economy. Penguin Classics*. London: Penguin Books, 1993.

Mindell, David A. *Our Robots, Ourselves: Robotics and the Myths of Autonomy*. New York: Penguin, 2015.

Moulier Boutang, Yann. *Cognitive Capitalism*. Cambridge: Polity Press, 2011.

Naastepad, C. W. M., and Jesse M. Mulder. 'Robots and Us: Towards an Economics of the "Good Life."' *Review of Social Economy* 6764 (2018): 1–33. https://doi.org/10.1080/00346764.2018.1432884.

Powell, Benjamin. *Out of Poverty: Sweatshops in the Global Economy. Cambridge Studies in Economics, Choice, and Society*. New York: Cambridge University Press, 2014.

Schneider, Nathan, and Jeanelle Orsi. '5 Ways to Take Back Tech.' *TheNation.com*, 2015. https://www.thenation.com/article/archive/5-ways-take-back-tech/.

Scholz, Trebor. 'Platform Cooperativism vs. the Sharing Economy.' *Medium.com*, 2014. https://medium.com/@trebors/platform-cooperativism-vs-the-sharing-economy-2ea737f1b5ad.

Smythe, Dallas Walker. 'Communications: Blindspot of Western Marxism.' *Canadian Journal of Political and Social Theory/Revue Canadienne de Théorie Politique et Sociale* 1, no. 3 (1977). https://journals.uvic.ca/index.php/ctheory/article/view/13715.

Stilgoe, Jack. 'Machine Learning, Social Learning and the Governance of Self-Driving Cars.' *Social Studies of Science* 48, no. 1 (2017): 25–56. https://doi.org/10.1177/0306312717741687.

Terranova, Tiziana. 'Free Labor.' *Social Text* 18, no. 2 (2000): 33–58. https://doi.org/10.1215/01642472-18-2_63-33.

Vercellone, Carlo. 'The Crisis of the Law of Value and the Becoming-Rent of Profit: Notes on the Systemic Crisis of Cognitive Capitalism.' In *Crisis in the Global Economy: Financial Markets, Social Struggles, and New Political Scenarios*, edited by Andrea Fumagalli and Sandro Mezzadra, 85–118. The MIT Press, 2010.

———. 'From the Crisis to the "Welfare of the Common" as a New Mode of Production.' *Theory, Culture & Society* 32, no. 7–8 (1 December 2015): 85–99. https://doi.org/10.1177/0263276415597770.

———. 'From Formal Subsumption to General Intellect: Elements for a Marxist Reading of the Thesis of Cognitive Capitalism.' *Historical Materialism* 15, no. 1 (2007): 13–36. https://doi.org/10.1163/156920607X171681.

Vrousalis, Nicholas. 'How Exploiters Dominate.' *Review of Social Economy* 79, no. 1 (2019): 103–30. https://doi.org/10.1080/00346764.2019.1618483.

Zuboff, Shoshana. 'Surveillance Capitalism and the Challenge of Collective Action.' *New Labor Forum* 28, no. 1 (24 January 2019): 10–29. https://doi.org/10.1177/1095796018819461.

———. *The Age of Surveillance Capitalism: the Fight for a Human Future at the New Frontiers of Power*. London: Profile Books, 2019.

6 AI and good work in a pluralistic society

6.1 The Wrench and the lead

> … like I told you, it's no accident that I'm in this line of work, going from one construction site to another, to all the factories and ports of the world: it's what I wanted. All kids dream of going into the jungle or the desert or Malaya, and I also had those dreams, only I like to have my dreams come true … There were two ways for me: I could wait till I got rich and then be a tourist, or else I could be a rigger. So I became a rigger. … I like to visit foreign countries … By now, it's become such a habit that if I had to settle down quietly, I'd get sick. If you ask me, the world is beautiful because it's all different.[1] (Levi 2017, 2)[2]

With these words, Tino Faussone, the main character in Primo Levi's 1978 novel *The Wrench*, explains his decision to become a rigger. He wanted to travel the world. Of course, other people may have different reasons for becoming or not becoming riggers. As Faussone says, the world is beautiful because it's all different, or, in more philosophical terms, we live in a pluralistic society where different legitimate conceptions of the good are present. Works are all different, and so are people. What makes, say, nursing good (to nurses and society) is different than what makes constructing good (to construction workers and society), what makes teaching good (to teachers and society), etc. Moreover, individual people may have different reasons to value the same work. Also, in addition to the individual perspective, there is a collective one: 'goods of work' are also the 'constitutive goods' achieved by different professions from a societal point of view, which makes construction, healthcare, defence, research, etc. good for society and not only for the workers.

Consider the story of LeeAnne Walker, a mother of four in Flint, Michigan, in the 2010s. After her children got seriously sick due to an excessive level of lead in the water of the city, she decided to learn from scratch the science and technique of water sample testing to be able to

DOI: 10.4324/9781003303244-8

prove, with the help of a professional scientist and many other citizens-turned-scientists, that people in Flint had been exposed to lead poisoning by the decision of the local authorities to temporarily pump water to the house from a highly contaminated river.[3] Scientific work, here performed by a new and original combination of professional and non-professional scientists, allowed people in Flint to uncover political corruption and protect their health. Science, like many other professions, also has a societal value, and thus there are also societal reasons to protect its constitutive goods from being affected by technological change or, like in the case of Flint, to organise scientific and technological change in such a way that those goods are protected and enhanced.

In this chapter, I address the question of to what extent the introduction of artificial intelligence (henceforth AI) and digital technologies can make work better or worse. This question, as I argue, requires a complex answer that takes seriously the legitimate differences in the perspectives of different workers and what I will call the different 'constitutive goods' of different work domains from a societal point of view.

I will first address the worker's perspective, followed by the broader societal perspective. From the worker's perspective, I will challenge the idea of designing technology to make work better for individuals with the quality of work being defined from an alleged objective and universal perspective: the 'goods of work' typical of mainstream (Western) moral philosophy.[4] One typical illustration of this approach is the proposal to design technology to automate and replace 'dirty, dull, and dangerous jobs,' i.e. jobs that allegedly do not bring any positive good to the workers. As I will explain, this is a morally problematic goal insofar as it may preclude some people from achieving what they consider to be, despite everything, their preferred work. Taking a more pluralistic perspective, I propose to focus on policies and design principles that would allow for a more just distribution of the opportunities to achieve good work. This could be achieved by including some 'goods of work' among the basic goods allocated in traditional (i.e. broadly Rawlsian) theories of distributive justice. The idea is to design technology and institutions to allow everyone a fair opportunity to achieve at least some of the goods they may attach to different jobs and tasks.

The second part of this chapter addresses a related but different kind of issue, that is, to what extent the introduction of AI and new digital technologies more generally may create a shift in the 'constitutive goods' of specific work domains, thereby making it more difficult or easier for workers and society to achieve these goods, for instance, by changing the nature of the defence profession or scientific (healthcare) research. I will provide a framework that can help better understand and address this question based on Michael Walzer's idea of 'spheres of justice.'[5]

The last part of this chapter illustrates this approach with some examples of shifts of goods related to introducing new technologies, two undesirable: the depoliticisation of war and the 'Googlisation' of healthcare research, and one desirable: the democratisation of scientific research.

6.2 The goods of work for workers

6.2.1 Good work in Western philosophy

A long trend in Western philosophy has converged in identifying some key elements that make work good. According to this tradition, work is good, generally speaking, insofar as it allows persons to develop their mental and physical capacities and maintain a meaningful relationship with the object of their work – be that a physical object or a process – and with other people. This idea received different and partially overlapping interpretations, with, for instance, Aristotle stressing the importance of developing complex personal and social skills (virtues), Kant that of developing one's autonomous agency, Hegel's that of work as building up one's social identity and creating social recognition, and Marx that of remaining connected in the right way with some human essential features, with the product of one's work and with the other human beings involved in the production system, that is, in Marxian terms, not being alienated.

Recently, philosophers Anca Gheaus and Lisa Herzog synthesised and expanded this philosophical tradition and connected it with recent empirical findings supporting it.[6] They proposed the following four sets of 'goods of work' (other than money) to be added to the absence of 'bads' like physical and psychological harm, exploitative salary, etc.:

(1) 'Excellence': the Aristotelian idea that good work produces knowledge, technological achievements, or beauty. Good work develops one's (mental) capacities and skills. Through the development of these capacities, some valuable results are produced such as, for instance, a piece of craftsmanship or a sound medical diagnosis. By also including in the capacities to be developed that of formulating, revising, and pursuing their concept of the good, it is possible to also have the Kantian ideal of moral autonomy as a good to be achieved through work.[7]

(2) 'Social contribution': contributing to some socially valuable achievement such as giving people material goods or services, helping others achieve their goals or improving their well-being. It is the common idea that work gives us a purpose, or in the words of the psychologist Studs Terkel, 'daily meaning as well as daily bread.'[8] Indirect evidence of the importance of social contribution is that work considered socially relevant – education, healthcare, assistance to people experiencing poverty, etc. – is often done for insufficient salary if not *pro bono*.

(3) ‘Community’: doing things together to build meaningful relationships with colleagues or co-workers. Gheaus and Herzog interpret this in a strongly egalitarian way and claim that work relationships are good only if they are ‘free and equal’;[9] they endorse the idea that there should be some democratic structures in the workplace, also because ‘long-time exposure to very competitive or hierarchical relationships undermines individuals’ psychological ability to form and sustain egalitarian relationships.’[10] A less committal view is that some meaningful, non-exploitative, social relationships are part of good work.

(4) ‘Social recognition’: the Hegelian idea that through professional identity, it is possible to get social recognition, that is, being seen as someone with a recognisable place in society, a status. This also contributes to shaping one’s identity. A person’s social status is influenced by the work they do and their hierarchical position within their workplace. This point is widely supported by shared experience and empirical research in social science. However, it is also important to notice that things often work the other way around: the social status outside work – as determined by independent factors such as (family) income, education, gender, race, etc. – influences the possibility and probability to achieve a professional position and the corresponding status.[11]

6.2.2 *Giving bad work to machines: a misplaced philosophical ideal?*

As Western philosophers or people educated in mainstream Western culture, it is tempting to endorse the list of goods as described by Gheaus and Herzog, which nicely synthesised some fundamental moral ideals about work as developed in our culture. According to empirical findings in (Western) social science, these moral ideals have been endorsed by many people. Moreover, it is tempting to use this list as a compass to guide the assessment, regulation, and design of AI technology in the workplace.

For instance, a recurring claim in the debate on the ethics of AI and robotics is that a legitimate and commendable societal goal is the design of AI systems or robots that would replace jobs or tasks that are dirty, dull, physically challenging, lonely, repetitive, or not very rich in goods of work (of the Western philosophy), such as cleaning, mining, serving, cashing, storing, delivering, etc.

Consider the following example taken from the healthcare domain: a robot that collects urine samples in a hospital by entering the patient’s rooms to protect the health of the care professional and the privacy of patients while at the same time nurses remain present, engaged, and responsible for the overall process, the communication and trust relation with the patients, and for supervising the delivery of the samples to the laboratory.[12] At first sight, this looks like a good solution – robots taking over a so-called dirty, dull, and dangerous task (urine collection) and people being able to focus on more rewarding tasks that typically require human capacities (taking responsibility, interacting with other professionals, and taking care of patients). However, things are more complex than they seem. In a pioneering 1983 paper, eloquently titled *Ironies of Automation*, cognitive scientist Lisanne Bainbridge discussed how automation of work processes 'may expand rather than eliminate problems with the human operator.'[13] This happens because the more advanced the use of technology in sensitive domains, the more crucial the contribution of human operators in supervising and preventing accidents. Recent studies in the so-called human factor science – the psychology of the human interaction with machines – somehow confirm that creating 'teams'[14] of humans and automated agents is a challenging task, one that may create specific extra tasks and psychological burdens for the human agents, typically those involved in the supervision of the automated agents and the overall coordination of the new team.[15] As an illustration of this phenomenon, think of one of the domains where automation has had the most remarkable results: civil aviation. Modern aeroplanes are equipped with autopilot technologies that can guide the aeroplanes for long times and under challenging circumstances. Despite this, or rather *because* of this, a team of highly qualified professionals should constantly supervise the behaviour of the system, both on the plane and from the ground, to prevent risks and accidents, and ensure the smooth functioning of the system. Similar scenarios can be imagined, *mutatis mutandis,* in other circumstances where people's vital interests are put in the hands of not only transportation professionals, but also, for instance, healthcare, financial and legal professionals.

Moreover, even if automation would relieve human workers from repetitive or menial tasks, making them able to focus entirely on those tasks that require their specific professional human skills – in the case of care professionals: attentiveness, empathy, responsibility, etc. – still the quality of their work may not improve. In fact, as the philosopher of technology Shannon Vallor has remarked, professionals would now find themselves performing intense emotional and relational labour for longer periods, potentially for all their working hours, because they no longer would have to do those activities that used to give them a mental break.[16] The alleged freedom from the burden of meaningless or not-so-human tasks may burden them with too many demanding, energy-consuming, and often stressful and burdensome, all-too-human tasks. This results in the psychological challenge of an overload of care, meaning, and engagement (or, depending on the profession, any other alleged human good or 'excellence'). The *hyper-humanisation of work* would be a new and interesting twenty-first-century 'irony of automation.'

Of course, things need not necessarily play out in this way. Back in the 1930s, the Nobel prize laureate economist John Maynard Keynes, at the end of his career, imagined a future society where, thanks to the increased productivity created by machines, people could work less and enjoy more free time.[17] So, it is in principle imaginable that, for instance, care workers will get shorter work times to compensate for higher engagement in psychologically draining tasks. We might get there one day. However, almost one century after Keynes' speculation and after a significant increase in technology-driven productivity, social and economic systems in industrialised countries still do not seem to offer many opportunities to achieve a corresponding reduction in working time. More productivity has not yet allowed for a corresponding reduction of work (at the same salary).

There is another reason to be sceptical about the idea of striving to eliminate jobs or tasks that mainstream Western moral philosophy does not consider worthwhile. From a more principled perspective, some people may, under some conditions, have good reasons to enjoy that kind of work or prefer it to other jobs (or to no job at all). Simple, repetitive, or otherwise so-called 'automatable' tasks may fit their psychological profile or their physical needs (some people may enjoy repetitive or physically challenging and even dangerous tasks), or the job may fit well in some broader life plan of theirs, for instance, a young person looking for a side-job to make some money while studying or someone looking for a simple job that leaves them time and energy to pursue other activities outside working hours (sport, music, volunteering, and care of family). Just eliminating jobs or tasks that do not look good from some alleged universal moral point of view may deprive some people of their preferred options to achieve their good work (and life).

The general lesson we should draw is that we cannot take any of the goods of work – defined in some general, objective terms – as a *necessary* condition for achieving good work. People's views about what constitutes good work may and do legitimately diverge. Some people may reasonably prefer a job that gives them good social connections and social contribution without it being particularly challenging; others may value the social connections and not even care about whether their job has a recognisable social contribution; others may prefer a job that offers an excellent intellectual challenge while not having an immediate social contribution and instead making them socially relatively isolated, e.g. scientific or philosophical research on very abstract or obscure subjects; and yet others may want a non-challenging job that pays the bills and leaves them free to enjoy life outside the workplace. Recognising the existence and legitimacy of these differences is important to avoid the risks of what in moral philosophy is called *perfectionism and paternalism*, the idea that even if people think they live a good life and do so according to some recognisable standards, they still should strive to improve their standards (perfectionism), where the improvement is defined according to a broader list of requirements they may not endorse, or even strongly oppose. If they cannot or do not want to, they should be encouraged and even pushed to do so, in their alleged best interest (paternalism). In the example above, the person happily employed in a company that gives them simple and repetitive tasks, and happily pursuing what they consider to be a good life should be incentivised to train and develop to be able to get a different, arguably more meaningful work, even though they do not want to, and do not believe they should. Perfectionism and paternalism are incompatible with a free and plural society, where people are allowed and encouraged to pursue different conceptions of the good life and good work.[18] Also, from a global perspective, many forms of perfectionism and paternalism developed within Western culture may betray some ethnocentrism, the idea that – in this case – some Western ideals should be pursued globally.

6.2.3 Distributing the goods of works more justly

On the other hand, it has been argued that *not* pursuing any political – and I will add: design – action to support the realisation of the goods of work of the (Western) moral tradition would have an unacceptable impact on people's freedom. Unregulated labour markets are likely to be dominated by the interests of capital owners and managers: enhancing productivity and efficiency, often at the expense of workers' capacity to enjoy other goods of work. So, people who want to achieve some of these goods will be *de facto* prevented from doing so. [19]

This should be a paramount political concern since it is very likely that many who are prevented from achieving these goods inside the workplace are also prevented from pursuing them outside. In fact, in the current social and economic conditions, in the absence of social schemes that allow one to make one's living without working (like a high Universal Basic Income), work activities in the economic sense – that is, paid jobs – take most of adult life for most people. For many people, the possibility of achieving a good life crucially depends on the possibility of having a good job ('good' in the broader sense discussed here). Moreover, many elements of the good life cannot be bought to be enjoyed outside working hours, so the idea of just using work to afford a good life in one's free time is challenging to realise for many.[20] Also, the conditions of work often affect life outside work, not only by determining people's economic conditions but also by giving them a specific culture, status, identity, and position in a social network that will affect the quality of their life and the possibility to achieve their goals.[21]

Therefore, one may endorse a so-called 'mild perfectionist' approach. 'Perfectionist' as it does endorse a specific view about the goods of work – the one captured in the mainstream Western moral philosophical perspective on work; 'mild' because it proposes that policies are put in place to only support people who want to pursue this view, as opposed to supporting this view for everybody, thereby preventing other conceptions of good work from being realised. One possible implication of such a mild perfectionist approach is that ensuring fair access to the goods of work should be part of distributive justice; that is, the goods of work should be considered as 'primary' or 'basic' goods the distribution of which should be regulated by the State (like income, access to essential services, etc.).[22]

One may develop this idea, for instance, by expanding a Rawlsian approach to distributive justice to include the goods of work. Famously, John Rawls aimed to create a theory of justice acceptable in a pluralistic and multicultural context, where people hold different conceptions of the good. Broadly speaking, Rawls' framework sees justice as an ideal concerning the distribution of resources. Relevant resources whose distribution should be regulated by a theory of justice are called 'primary goods' by Rawls – that is, 'all-purpose means' such as income, wealth, or what he calls 'the social basis of self-respect.' Rawls wonders what would be a distribution of these goods that people should accept, again, without considering their specific (moral) identity or value preferences, or, in Rawls' terms, if they were to choose without knowing their position in society: their age, gender, income, level of education, moral, or religious orientation and others. This is the famous 'original position' under the 'veil of ignorance.' According to Rawls, under these ideal conditions, rational people would only accept differences in the distribution of primary goods that (a) are generated in a context of equal opportunity of access, as they do not want to be systematically penalised whoever they turn out to be, and that (b) raise the level of the least advantaged in society, as they want to get support in case they turn out to be among the worst-off ones in a society that allows for socio-economic differences (the 'difference principle').[23]

By applying Rawls' theory to the distribution of the goods of work, one could obtain the following principles. Differences in the achievement of the goods of work may be justified only to the extent that, firstly, specific tasks are not assigned according to features such as gender, income, race, etc., and there is a fair opportunity for people from any social group (not) to choose them. This is an important requirement since the opposite is very often the case: the phenomenon of 'job segregation' where people from certain groups (gender, ethnicity, nationality, etc.) are pushed towards specific jobs and not allowed to access others. In other words, the competition to get or avoid these jobs is not fair.[24] Secondly, those who end up doing jobs that are in some way less desirable are put in the position to achieve some other goods (other than money and decent working conditions) in their work if they want. For instance, psychologists Ashforth and Kreiner showed that persons doing the so-called 'dirty work' (e.g. cleaning) often develop strong social bonds and a common culture at work through which they can build a recognisable social identity that supports their self-esteem against the social stigma usually associated to their work. [25]

From a technological point of view, this suggests a different approach; rather than only investing in technologies that will (possibly, one day) fully replace jobs or tasks that look undesirable according to some – allegedly objective – universal perspective, we should also wonder how to design technological systems that will make these jobs or tasks good or better for those who end up doing them. 'Good' means right-respecting, sufficiently paid, not too physically and psychologically taxing as well as supporting the achievement of some of the 'goods' that people may strive to achieve by doing them, for instance, social bonds and social recognition for a physical or psychological challenging job exposed to social stigma like cleaning.

6.3 Shifts of goods

The previous sections have focused on what it means to achieve the 'goods of work' from the perspective of individual workers and what principles should guide political and technological interventions that may promote or at least not hinder the achievement of these goods by them. I have endorsed a pluralistic and (distributive) justice-based approach. Pluralistic as it recognises that people may legitimately be willing to pursue different goods in their work. Distributive justice-based as it recognises that it is not always possible for the job market, the State, or technological development to grant *everybody* the possibility to achieve their preferred job or set of tasks within a job. Still, the State is arguably obligated to give all people an equal opportunity to try and access the goods of work by providing access to education and skill acquisition and by removing social structures that may lead to job segregation (people being assigned jobs or tasks based on their gender, nationality, etc.). From a design perspective, the design of jobs, tasks, and technologies should allow people to achieve at least some of the goods of work they want to achieve.

However, this picture is not complete. Individual workers are just some of the ones with a legitimate interest in achieving the goods of work. Work should provide goods to other people in society as well. Remember that even if nobody found cleaning tasks good in any way, things still need to be cleaned. People other than workers have legitimate expectations towards achieving the goods of work. Healthcare should be good for healthcare professionals and patients, scientific research for researchers and society, and so on. Technological development may impact the freedom to achieve some goods of work in yet another way: by creating broad shifts of goods within individual professional domains, which may make it more difficult for people in society to benefit from some goods of that profession.

To address this angle of the issue, I propose to integrate the pluralist and distributive justice approach proposed so far with a complementary approach to social justice that has addressed the issue of the plurality of values differently: Michael Walzer's theory of justice as the 'art of separation' between 'spheres of justice.'

6.3.1 Spheres of Justice[26]

In *Liberalism and the Art of Separation*,[27] Michael Walzer claims that distributive justice is not just a matter of transferring wealth or other general resources from the wealthy to the poor; it is something different and more complex – in Walzer's words, a 'matter of getting the lines right'[28] and building political and institutional – I will add: socio-technological – structures to prevent unwanted influences between the different spheres of life. In *Spheres of Justice*,[29] Walzer further explains how these lines should be drawn. Whereas traditional liberals have primarily drawn the lines around individuals' activities and rights, Walzer proposes to look at the principles defining and governing different social domains and activities, which he calls spheres. Different societal spheres revolve around different goods – in a moral/social, not economic sense – and are defined by different constitutive aims and purposes. Health care is about medical care, politics is about public office, the market is about exchanging material goods, education is about learning, etc. These different spheres also have criteria, norms, and moral rules that govern the regulation, governance, and distribution of things that matter. Heart surgery is, for instance, usually given based on a need to those who need it most, public office to those who were democratically elected, money using some market mechanism for value-added and work done, etc. According to Walzer, in contrast to Rawls and even more Van Parijs and the Basic Income supporters, there are no universal principles and 'all-purpose' primary goods that can validly be applied across all spheres of life. It is, therefore, widely considered inappropriate to use criteria and normative rules from one societal sphere to guide behaviour decision-making and distribution in another sphere. Were that to happen, things that ought not to be mixed up would get mixed up. In the market, for example, it may be all right to compromise, bargain, and strike deals to some extent. In science, politics, education, and family life, our choices and decisions should arguably not be affected by this market logic rife with compromise and bargains. By the same reasoning, profit motives are considered inappropriate in health care. This is not the same as saying that cost and efficiency considerations have no role in health care or other sectors. According to Walzer, a just society exhibits 'complex equality,' i.e. a just society respects the moral logic of different social spheres, roles, tasks, and aims.

Therefore, the art of separation consists of avoiding normative cross-contamination or moral short circuits. For similar reasons, we should reject the idea that people can convert advantages in one sphere to advantages in another sphere, as well as the idea that possession of one type of good (e.g. health, money, and education) will also get you everything else (e.g. public office, university degrees, places on a waiting list, and love). This idea of social differentiation in delineated systems and subsystems and the relative autonomy of these (sub)systems can also be found in the work of sociologists such as Luhmann and Bourdieu. Moreover, Walzer's proposal to protect different spheres of justice partly resonates with Jürgen Habermas' criticism of modern capitalism as leading to the 'colonisation of the lifeworld.'[30] In Habermas' terms, money and power have colonised areas of social life that are dependent on communicative actions and social relations driven by different normative expectations.[31]

Walzer's perspective may provide a broader framework to deliberate about justified constraints and shared moral requirements for designing future technological systems that provide a plurality of goods to workers and others. As I argue, the fast development of AI may create shifts of goods within many professions; that is, some goods or values typically embedded in functioning of AI systems may end up replacing the goods of various professions. In Habermas' terms, data-driven AI logic may colonise the professional world. I will illustrate this trend by looking at two examples: the depoliticisation of warfare and the Googlisation of healthcare research. However, I will also conclude on a positive note. When designed to embed the appropriate set of goods, digital systems may also support the realisation of the goods of a given profession. This will be illustrated with the example of the democratisation of science.

6.3.2 The depoliticisation of warfare

Let us consider the automation and digitalisation of warfare operations made possible by AI and robotics. Some military personnel have complained that the deployment of drones and other remotely controlled weapon systems is depriving them of the possibility to develop what they see as the typical forms of excellence of their profession (and of their life) such as courage on the battlefield and deployment of physical skills. For some, military operations are about action on the battlefield, not mastering a joystick. In 2015, US Defence Secretary Chuck Hagel famously cancelled the combat medal for drone operators, claiming that combat medals should be reserved for those 'who incur the physical risk and hardship of combat.'[32] The medal had been dismissively called the 'Nintendo medal' and drone operators 'desk jockey warriors.'[33] While certainly an important concern to consider, we cannot just conclude *a priori* that AI military technologies should not threaten the traditional ways of waging war.

First, as it has been widely remarked, the (military) profession has changed across the centuries and so have the virtues and skills associated with it or their exemplification. Also, remote drone operators or, as they have sometimes been called, 'cubical warriors' are far from being careless videogamers: they are exposed to intense emotional and psychological pressure and they have been reported to suffer high levels of Post-Traumatic Stress Disorders.[34] And some soldiers may prefer the joystick to the action on the field anyway. As for the risks that professionals may not be able to practice valuable virtues outside their profession (courage, etc.), these virtues may remain available in other professions, say disaster management, so that if they are not available in the military profession anymore, people may still turn to these other professions to cultivate them.

Also, from a societal perspective, one equally if not more pressing issue is to what extent the development and use of technology may change not just the experience of war for military personnel but also, in a Walzerian spirit, the constitutive shared goods, principles, and values of the defence domain.

Technological development has always been a critical factor in changing international relations. The invention of the atomic bomb shaped the international politics of the second half of the twentieth century and still plays a crucial role in present-day international relations following the Russian invasion of Ukraine. Similarly, the availability of AI-based systems to States may also significantly impact international relations and defence policies. Much talk is about the impact of futuristic, fully automated warfare and other *Terminator*-like scenarios, but some important shifts have already happened. Since the Obama administration, for instance, the US government has heavily shifted from traditional war operations to drone-operated targeted killings and attacks.[35] Over the past years, China and Russia have allegedly made intense use of cyberattacks or so-called 'cognitive warfare.'[36] While one may rightfully point out that there are no such things as 'goods' of war – war is terrible anyway – it is undoubtedly the case that the introduction of advanced digital and AI technology may shift the principles and methods of warfare. Whether this is a change for the better or the worse critically depends on our understanding and interpretation of these principles. Governments and military commanders may deem it reasonable for strategic, defence, and security objectives to be achieved more effectively and with less deployment of human personnel on the ground. However, one may argue that these goals should also be, at least to some extent, part of an open, democratic and transparent deliberation process, and the invisibilisation of war may bring too much 'depoliticisation' and 'intelligentisation' of defence security policies and operations instead. The risk is not only that AI and robotics will lower the threshold of waging war,[37] or leading to an increase of war crimes, but also that the gradual transforming of war into something different, which is less and less dependent on public political decisions and overt operations and more and more on covered operations decided by alleged experts.

In the Walzerian spirit of 'spheres of justice,' one may say that there is a risk of unwanted cross-contamination between the logic of (just) war, which must crucially involve some engagement with political and legal principles of justice and accountability and a pure logic of national security and strategic advantage. Paraphrasing Saint Augustin, justice being taken away States are nothing but big criminal gangs.[38] This is not the place to present, let alone take a final position on a debate that is much richer and more articulated than this section allows for.[39] From the point of view of the goods of work, the point is that technological change may impact the military profession not only by affecting the possibility for personnel to develop and practice what they may consider to be some important features of their work, but also by shifting what *the society considers* to be important principles, 'goods' of that profession.

6.3.3 The Googlisation of (health) research

Another domain in which AI will be more and more deployed is healthcare. Healthcare is a broad concept, covering various activities ranging from diagnosing and surgery to assistance and companionship. Here too one can distinguish between the impact of technology from the perspective of the individual actors involved and from a broader societal perspective. One recurring concern with the introduction of new technologies in healthcare is the risk of dehumanisation. Depending on the activity involved, this risk may be more or less severe. Activities strongly characterised by their end goal, such as surgery, can arguably be crucially enhanced by AI and robotics technology without creating relevant risks of dehumanisation of care. In fact, high-precision technology can dramatically improve the quality of care for patients and surgeons.[40] Other strongly goal-directed medical activities such as diagnosing, while being open to significant improvement through machine learning techniques, still raise some issues in terms of the capacity of the human doctor to make sense of the process, their role in it, and the result of the diagnosis and therefore of their capacity to maintain their professional self-identity and to give meaningful explanations to their patients. This will be discussed in the next chapter in the context of so-called moral accountability gaps. The introduction of AI in healthcare activities more squarely centred on human relations brings a higher risk of affecting critical human virtues and values of individuals involved. One recurring example is the use of the so-called social robots for the assistance and companionship of elderly persons or persons with medical conditions. One possible approach to designing and using these robots would be designing them in a way that explicitly protects and enhances the central values and virtues of the human people involved, not just the (cost-)efficiency of the process. For instance, robots serve as a cue for an increased social interaction with human personnel rather than as a replacement of that personnel.[41]

Here again, a task-based, individual worker-oriented approach does not allow us to appreciate the broader, collective, and political side of the picture. Tamar Sharon has recently discussed the 'Googlization of healthcare research.'[42] Big tech companies such as Google, Apple, and Alphabet have entered the healthcare and biomedical research field thanks to their expertise in data collection and analysis and the vast amount of personal data in their possession. In addition to the well-known issues of data protection, this also raises two issues due to a possible shift in the values and principles that drive future research. Firstly, these companies have no experience in the healthcare sector and are likely to inject into healthcare research and practice a different set of principles, values, and methodologies. These may create tension between, for instance, on the one hand, the traditional methodologies of biomedical research and the principles of healthcare practices and, on the other hand, the generic logic of data analytics and the profit-based business model of the companies. Secondly, the presence and increasing power of such private companies may heavily and unduly influence the research and investment agenda in the healthcare and biomedical field. A similar trend can be seen in academic research in general, with 'big tech' offering more and more infrastructure to researchers and teachers, arguably leading to a shift in the guiding principles of universities, moving from independent culture producers to corporate business.[43] Of course, being independent does not mean being value-free or abstracted from any political context. Scientific and academic research has never been free from political influences. It has always reflected a given political context's specific values, ideals, and interests as well as the implicit ethical and political commitments of researchers and research institutions. However, again, these values and principles must remain open to a plural debate, sensitive to local contexts, and not be dominated by one specific set of principles or values, in this case, those of the technology providers meant to support research.

6.3.4 The democratisation of science

However, AI and data-driven technologies must not necessarily harm the values and 'goods' pursued by professions. For example, AI technologies also allow for a broader range of participation of non-scientists in scientific research. Thanks to recent computer and sensor technology advancements and the creation of cheap open-source micro-controller boards, individual laypersons can now measure 'whatever they can think of in their environment.'[44] They can log all their data and compare them with those of other people. Online communities have been created around hardware and research topics, like spectralworkbench.org, where the spectra taken of different light sources are shared.

In this respect, it has been suggested that we seem to be back in the early days of the scientific revolution, where 'any interested person with time to spare could look at the stars, gather data and make predictions,' and 'scientists were often also farmers, artists, or statesmen.'[45] While we do not want to go back to the pre-scientific revolution age, it may be worth exploiting technological innovation to involve more non-professional scientists in producing scientific knowledge. However, in many current so-called citizen science projects run by academic scientists, participants typically only provide their time or computer power; their role is limited to data collection or classification. Their intelligence or creativity is not sufficiently utilised; the project only offers minimal opportunities for personal growth,[46] especially since participants are rarely involved in the research design. This is arguably a missed opportunity in many respects. From a scientific perspective, the design of research projects may be improved by including citizens' local knowledge and local networks.[47] From a policy and engineering perspective, local governments may fruitfully use the input of their citizens to better govern cities, and citizen communities may come up with small-scale solutions for challenges such as energy or food production or waste recycling. From a political perspective, citizen science may sometimes be an antidote to political and institutional corruption as illustrated by the Flint water case.

Flint is a city in Michigan, USA, on the shores of the river of the same name. It is the birthplace of General Motors and home to other industries. The Flint River has historically been severely contaminated by waste disposal of local industries, the city's raw sewage, and leaching landfills. The economic crisis associated with the car industry decline dramatically impacted the local economy: today, Flint has a population of about 100,000 people, primarily African-American, of whom about 45% live below the poverty line.[48] In 2011, the Michigan governor appointed an emergency manager to address the city's financial deficit. In a move presented as reducing the city's financial deficit, the decision was taken to stop piping treated water from Detroit and start temporarily pumping water from the Flint River through an ageing piping system. While Flint people protested and reported anomalies in the look, taste, and odour of their tap water and experienced symptoms of what would turn out to be lead poisoning (rashes and hair loss), authorities would insist that all safety measures had been taken and the levels of lead were within the limits of tolerance.[49]

One of the Flint inhabitants was LeeAnne Walters, 37 years, mother of four sons, who were all affected by rashes, with one constantly ill and suffering from an apparent stop in his growth. She decided to study the science of how to gather water samples that would accurately measure risk; she found out that the city officials' sample collection had been done in a lacking way to conceal the lead problem and that no erosion control had been done on the pipelines. After reporting her concerns to the city officials to no avail, she contacted regional officers of the Environmental Protection Agency (EPA). One of them, Miguel Del Toral, an expert on the Lead and Copper Rule, immediately sensed the seriousness of the situation and put her in touch with Marc Edwards, an expert on erosion control at Virginia Tech, for further testing. Edwards' tests revealed that the water contained the highest lead level he had seen in his 25-year career. When local officials still tried to ditch the evidence and discredit the citizens and civil servants supporting them, Edwards organised a largely self-funded project called the Flint Water Study, letting citizens re-do the water sample collection. He sent 300 testing kits to a church in Flint and instructed citizens on how to use them. Citizens' tests revealed the absolute magnitude of the lead problem in Flint. Their efforts were backed by an investigative reporter and a paediatrician from Virginia Tech who confirmed through a systematic study the gravity of the health conditions in the area. The results were widely publicly communicated by them, eventually pushing the country to declare a public health emergency and address the issue.

Communities in Flint invented a citizen science project to fight corrupted science and politics. Technology and data collection played an important role here. In other historical times, LeeAnne Walker, an ordinary citizen with no scientific education, would not have had easy access to water control information and instructions on water testing. Some of the mobilisation and the public protests of the Flint citizens would have been less effective without social media. However, a key factor for the success of the mobilisation was a new institutional organisation around science that some call 'community-based participatory research' (CBPR) methods. This approach, mainly utilised in public health, tries to promote equity between academic and community members by promoting 'shared control and leadership' between them in all aspects of the research.[50] In fact, in the Flint case, the discoveries of the experts were made possible by the community mobilisation strategies that exposed the potential problem with water contamination and invited the collaboration with experts, critically contributing to the generation of the research question and motivating the research, and then by collecting the data to answer those questions.

Sometimes, changes in the practices of a discipline can better promote the achievement of its constitutive goods. Citizen science is one example of how technological innovation coupled with new appropriate institutional design and governance may contribute to the realisation of an important ideal of scientific research: its capacity to produce knowledge through methods that are transparent and open to constant criticism and revision, potentially accessible to investigation by anybody and therefore not easily captured by political and economic powers. The extent to which citizen science and other platform-based social activities promote freedom and equal participation in science crucially depends on how these platforms and activities are organised. Philip Mirowski, an expert in the history of economics and the politics of science, rightfully insists that to produce real knowledge, citizen scientists must undergo real training, 'not a drive-by blast of methodology lite.'[51] However, he notices that most of the existing citizen science consists of the public donating its unpaid work and data to privately owned online entities that digest it as 'big data.'[52] He also warns that citizen science may be another tool for the public sector to reduce investments in science. The Flint experience shows that another citizen science is possible and that, when properly designed and supported, citizen science projects may promote scientific research through more democratic participation and citizen involvement.

Notes

1 Primo Levi, *The Monkey's Wrench* (New York: Simon and Schuster, 2017), 2.
2 The novel's Italian title (*La Chiave a Stella*) was translated as *The Wrench* in Great Britain and *The Monkey's Wrench* in the US. While the quote is taken from the US edition of the book, I prefer the British translation of the title, which is more faithful to the original.
3 Anna Mariya Barrie-Jester, 'What Went Wrong In Flint,' *Fivethirtyeight*, 2016, https://fivethirtyeight.com/features/what-went-wrong-in-flint-water-crisis-michigan/.
4 Anca Gheaus and Lisa Herzog, 'The Goods of Work (Other than Money!),' *Journal of Social Philosophy* 47, no. 1 (1 March 2016): 70–89, https://doi.org/10.1111/josp.12140.
5 Michael Walzer, *Spheres of Justice: A Defense of Pluralism and Equality* (New York: Basic Books, 1983).
6 Gheaus and Herzog, 'The Goods of Work (Other than Money!).'
7 Adina Schwartz, 'Meaningful Work,' *Ethics* 92, no. 4 (July 1982): 634–46, https://doi.org/10.1086/292380.
8 Studs Terkel, *Working: People Talk about What They Do All Day and How They Feel about What They Do* (New York: The New Press, 1972).
9 Gheaus and Herzog, 'The Goods of Work (Other than Money!),' 76.
10 Gheaus and Herzog, 78.
11 Cecilia L. Ridgeway, 'Why Status Matters for Inequality,' *American Sociological Review* 79, no. 1 (February 2014): 1–16, https://doi.org/10.1177/0003122413515997.
12 Filippo Santoni de Sio and Aimee van Wynsberghe, 'When Should We Use Care Robots? The Nature-of-Activities Approach,' *Science and Engineering Ethics* 22, no. 6 (7 December 2016): 1745–60, https://doi.org/10.1007/s11948-015-9715-4.

13 Lisanne Bainbridge, 'Ironies of Automation,' *Automatica* 19, no. 6 (1 November 1983): 775–79, https://doi.org/10.1016/0005-1098(83)90046-8.
14 'Teams' in scare quotes as it is a controversial issue whether it is ontologically correct and ethically desirable to talk of collaboration with robots and artificial agents in the workplace. See Katie D. Evans, Scott A. Robbins, and Joanna J. Bryson, 'Do We Collaborate With What We Design?,' *Topics in Cognitive Science*, 2023, https://doi.org/10.1111/tops.12682.
15 Amy R. Pritchett, So Young Kim, and Karen M. Feigh, 'Modeling Human–Automation Function Allocation,' *Journal of Cognitive Engineering and Decision Making* 8, no. 1 (1 March 2014): 33–51, https://doi.org/10.1177/1555343413490944.
16 Vallor's remark is contained in the following Vox article by Sigal Samuel, 'You Can Buy a Robot to Keep Your Lonely Grandparents Company. Should You?,' *Vox*, 9 September 2020, https://www.vox.com/future-perfect/2020/9/9/21418390/robots-pandemic-loneliness-isolation-elderly-seniors.
17 John Maynard Keynes, 'Economic Possibilities for Our Grandchildren,' in John Maynard Keynes, *Essays in Persuasion* (New York: Harcourt Brace, 1930), 358–73.
18 Richard J. Arneson, 'Meaningful Work and Market Socialism,' *Ethics* 97, no. 3 (1987): 517–45.
19 Gheaus and Herzog, 'The Goods of Work (Other than Money!)'; Richard J. Arneson, 'Meaningful Work and Market Socialism Revisited,' *Analyse & Kritik* 1 (2009): 139–51.
20 Gheaus and Herzog, 'The Goods of Work (Other than Money!).'
21 Denise Celentano, 'Work Justice Beyond Redistribution and Recognition,' in *Paradigms of Justice: Redistribution, Recognition, and Beyond*, eds. Denise Celentano and Luigi Caranti, 90–117 (London and New York: Routledge, 2021).
22 Gheaus and Herzog, 'The Goods of Work (Other than Money!).'
23 John Rawls, *A Theory of Justice* (Cambridge, MA: Harvard University Press, 1971).
24 Celentano, 'Work Justice Beyond Redistribution and Recognition.'
25 Blake E. Ashforth and Glen E. Kreiner, 'How Can You Do It?: Dirty Work and the Challenge of Constructing a Positive Identity,' *Academy of Management Review* 24, no. 3 (1999): 413–34, https://doi.org/10.5465/AMR.1999.2202129.
26 Credit is due to Jeroen van den Hoven for this section, which is an elaboration of a section of the following article: Filippo Santoni de Sio, Txai Almeida, and Jeroen van Den Hoven, 'The Future of Work: Freedom, Justice and Capital in the Age of Artificial Intelligence,' *Critical Review of International Social and Political Philosophy*, 13 December 2021, 1–25, https://doi.org/10.1080/13698230.2021.2008204.
27 Walzer, *Spheres of Justice: A Defense of Pluralism and Equality*.
28 Michael Walzer, 'Liberalism and the Art of Separation,' *Political Theory* 12, no. 3 (1984): 323.
29 Walzer, *Spheres of Justice: A Defense of Pluralism and Equality*.
30 Jürgen Habermas, *The Theory of Communicative Action. Vol. II: Lifeworld and System* (Boston: Beacon, 1987).
31 Timo Jütten, 'The Colonization Thesis: Habermas on Reification,' *International Journal of Philosophical Studies* 19, no. 5 (December 2011): 701–27, https://doi.org/10.1080/09672559.2011.629672.
32 Jesse Kirkpatrick, 'Drones and the Martial Virtue Courage,' *Journal of Military Ethics* 14, no. 3–4 (2 October 2015): 208, https://doi.org/10.1080/15027570.2015.1106744.

33 Robert James Sparrow, 'War without Virtue?,' in *Killing by Remote Control: The Ethics of Unmanned Military*, ed. Bradley Jay Strawser (New York: Oxford University Press, 2013), 84–105.
34 Kirkpatrick, 'Drones and the Martial Virtue Courage.'
35 Jessica Purkiss and Jack Serle, 'Obama's Covert Drone War in Numbers: Ten Times More Strikes than Bush,' *The Bureau of Investigative Journalism*, 17 January 2017, https://www.thebureauinvestigates.com/stories/2017-01-17/obamas-covert-drone-war-in-numbers-ten-times-more-strikes-than-bush.
36 Seumas Miller, 'Cognitive Warfare: An Ethical Analysis,' *Ethics and Information Technology* 25, no. 3 (4 September 2023): 46, https://doi.org/10.1007/s10676-023-09717-7.
37 Noel Sharkey, 'Killing Made Easy,' in *Robot Ethics: The Ethical and Social Implications of Robotics*, eds. Patrick Lin, Keith Abney, and George A. Bekey (Cambridge, MA: The MIT Press, 2012), 111–28.
38 'Remota itaque iustitia quid sunt regna nisi magna latrocinia?' wondered Saint Augustine in *De Civitate Dei*, Book IV, 4.
39 Accountability gaps are also discussed in Chapter 7 below.
40 Santoni de Sio and van Wynsberghe, 'When Should We Use Care Robots?'
41 Lillian Hung et al., 'The Benefits of and Barriers to Using a Social Robot PARO in Care Settings: A Scoping Review,' *BMC Geriatrics* 19, no. 1 (23 August 2019): 232, https://doi.org/10.1186/s12877-019-1244-6.
42 Tamar Sharon, 'The Googlization of Health Research: From Disruptive Innovation to Disruptive Ethics,' *Personalized Medicine* 13, no. 6 (November 2016): 563–74, https://doi.org/10.2217/pme-2016-0057.
43 Tobias Fiebig et al., 'Heads in the Clouds: Measuring the Implications of Universities Migrating to Public Clouds,' *ArXiv* abs/2104.09462 (2021), https://api.semanticscholar.org/CorpusID:261076516.
44 Diana Wildschut, 'The Need for Citizen Science in the Transition to a Sustainable Peer-to-Peer-Society,' *Futures* 91 (1 August 2017): 46–52, https://doi.org/10.1016/j.futures.2016.11.010.
45 Wildschut.
46 Dana Rotman et al., 'Dynamic Changes in Motivation in Collaborative Citizen-Science Projects,' in *Proceedings of the ACM Conference on Computer Supported Cooperative Work, CSCW*, 2012, 217–26, https://doi.org/10.1145/2145204.2145238.
47 Ângela Guimarães Pereira and Andrea Saltelli, 'Post-Normal Institutional Identities: Quality Assurance, Reflexivity and Ethos of Care,' *Futures* 91 (1 August 2017): 53–61, https://doi.org/10.1016/j.futures.2016.11.009; D. M. Warren, 'Using Indigenous Knowledge in Agricultural Development,' *World Bank Discussion Papers* 127 (1991), https://ideas.repec.org/p/fth/wobadi/127.html; Frank Fischer, *Citizens, Experts, and the Environment: The Politics of Local Knowledge*, (Durham, North Carolina: Duke University Press, 2002).
48 Melissa Denchak, 'Flint Water Crisis: Everything You Need to Know,' *NRDC* (blog) 8 November 2018, https://www.nrdc.org/stories/flint-water-crisis-everything-you-need-know.
49 Jennifer S. Carrera and Kent Key, 'Troubling Heroes: Reframing the Environmental Justice Contributions of the Flint Water Crisis,' *WIREs. Water* 8, no. 4 (2021): e1524, https://doi.org/10.1002/wat2.1524.
50 Carrera and Key.
51 Philip Mirowski, 'Against Citizen Science,' *Aeon*, 2017, https://aeon.co/essays/is-grassroots-citizen-science-a-front-for-big-business.
52 Something similar to what happens on other mainstream internet platforms, see Chapter 5.

References

Arneson, Richard J. 'Meaningful Work and Market Socialism.' 97, no. 3 (1987): 517–45.

———. 'Meaningful Work and Market Socialism Revisited.' *Analyse & Kritik* 1 (2009): 139–51.

Ashforth, Blake E., and Glen E. Kreiner. 'How Can You Do It?: Dirty Work and the Challenge of Constructing a Positive Identity.' *Academy of Management Review* 24, no. 3 (1999): 413–34. https://doi.org/10.5465/AMR.1999.2202129.

Bainbridge, Lisanne. 'Ironies of Automation.' *Automatica* 19, no. 6 (1 November 1983): 775–79. https://doi.org/10.1016/0005-1098(83)90046-8.

Barrie-Jester, Anna Mariya. 'What Went Wrong in Flint.' *Fivethirtyeight*, 2016. https://fivethirtyeight.com/features/what-went-wrong-in-flint-water-crisis-michigan/.

Carrera, Jennifer S., and Kent Key. 'Troubling Heroes: Reframing the Environmental Justice Contributions of the Flint Water Crisis.' *WIREs. Water* 8, no. 4 (2021): e1524. https://doi.org/10.1002/wat2.1524.

Celentano, Denise. 'Work Justice Beyond Redistribution and Recognition.' In *Paradigms of Justice*, edited by Denise Celentano and Luigi Caranti, 90–117. London and New York: Routledge, 2020.

Denchak, Melissa. 'Flint Water Crisis: Everything You Need to Know.' *NRDC* (blog), 8 November 2018. https://www.nrdc.org/stories/flint-water-crisis-everything-you-need-know.

Evans, Katie D., Scott A. Robbins, and Joanna J. Bryson. 'Do We Collaborate With What We Design?' *Topics in Cognitive Science* (2023). https://doi.org/10.1111/tops.12682.

Fiebig, Tobias, Seda F. Gürses, Carlos Hernandez Gañán, Erna Kotkamp, Fernando Kuipers, Martina Lindorfer, Menghua Prisse, and Taritha Sari. 'Heads in the Clouds: Measuring the Implications of Universities Migrating to Public Clouds.' *ArXiv* abs/2104.09462 (2021). https://api.semanticscholar.org/CorpusID:261076516.

Frank, Fischer. *Citizens, Experts, and the Environment: The Politics of Local Knowledge*. Durham, NC: Duke University Press, 2002.

Gheaus, Anca, and Lisa Herzog. 'The Goods of Work (Other than Money!).' *Journal of Social Philosophy* 47, no. 1 (1 March 2016): 70–89. https://doi.org/10.1111/josp.12140.

Guimarães Pereira, Ângela, and Andrea Saltelli. 'Post-Normal Institutional Identities: Quality Assurance, Reflexivity and Ethos of Care.' *Futures* 91 (1 August 2017): 53–61. https://doi.org/10.1016/j.futures.2016.11.009.

Habermas, Jürgen. *The Theory of Communicative Action. Vol. II: Lifeworld and System*. Boston: Beacon, 1987.

Hung, Lillian, Cindy Liu, Evan Woldum, Andy Au-Yeung, Annette Berndt, Christine Wallsworth, Neil Horne, Mario Gregorio, Jim Mann, and Habib Chaudhury. 'The Benefits of and Barriers to Using a Social Robot PARO in Care Settings: A Scoping Review.' *BMC Geriatrics* 19, no. 1 (23 August 2019): 232. https://doi.org/10.1186/s12877-019-1244-6.

Jütten, Timo. 'The Colonization Thesis: Habermas on Reification.' *International Journal of Philosophical Studies* 19, no. 5 (December 2011): 701–27. https://doi.org/10.1080/09672559.2011.629672.

Keynes, John Maynard. 'Economic Possibilities for Our Grandchildren.' In John Maynard Keynes, *Essays in Persuasion*, 358–73. New York: Harcourt Brace, 1930.

Kirkpatrick, Jesse. 'Drones and the Martial Virtue Courage.' *Journal of Military Ethics* 14, no. 3–4 (2 October 2015): 202–19. https://doi.org/10.1080/15027570.2015.1106744.

Levi, Primo. *The Monkey's Wrench*. New York: Simon and Schuster, 2017.

Miller, Seumas. 'Cognitive Warfare: An Ethical Analysis.' *Ethics and Information Technology* 25, no. 3 (4 September 2023): 46. https://doi.org/10.1007/s10676-023-09717-7.

Mirowski, Philip. 'Against Citizen Science.' *Aeon*, 2017. https://aeon.co/essays/is-grassroots-citizen-science-a-front-for-big-business.

Pritchett, Amy R., So Young Kim, and Karen M. Feigh. 'Modeling Human–Automation Function Allocation.' *Journal of Cognitive Engineering and Decision Making* 8, no. 1 (1 March 2014): 33–51. https://doi.org/10.1177/1555343413490944.

Purkiss, Jessica, and Jack Serle. 'Obama's Covert Drone War in Numbers: Ten Times More Strikes than Bush.' *The Bureau of Investigative Journalism* (17 January 2017). https://www.thebureauinvestigates.com/stories/2017-01-17/obamas-covert-drone-war-in-numbers-ten-times-more-strikes-than-bush.

Rawls, John. *A Theory of Justice*. Cambridge, MA: Harvard University Press, 1971.

Ridgeway, Cecilia L. 'Why Status Matters for Inequality.' *American Sociological Review* 79, no. 1 (February 2014): 1–16. https://doi.org/10.1177/0003122413515997.

Rotman, Dana, Jenny Preece, Jen Hammock, Kezee Procita, Derek Hansen, Cynthia Parr, Darcy Lewis, and David Jacobs. 'Dynamic Changes in Motivation in Collaborative Citizen-Science Projects.' In *Proceedings of the ACM Conference on Computer Supported Cooperative Work, CSCW*, 217–26, 2012. https://doi.org/10.1145/2145204.2145238.

Samuel, Sigal. 'You Can Buy a Robot to Keep Your Lonely Grandparents Company. Should You?' *Vox*, 9 September 2020. https://www.vox.com/future-perfect/2020/9/9/21418390/robots-pandemic-loneliness-isolation-elderly-seniors.

Santoni de Sio, Filippo, Txai Almeida, and Jeroen van den Hoven. 'The Future of Work: Freedom, Justice and Capital in the Age of Artificial Intelligence.' *Critical Review of International Social and Political Philosophy* (13 December 2021): 1–25. https://doi.org/10.1080/13698230.2021.2008204.

Santoni de Sio, Filippo, and Aimee van Wynsberghe. 'When Should We Use Care Robots? The Nature-of-Activities Approach.' *Science and Engineering Ethics* 22, no. 6 (7 December 2016): 1745–60. https://doi.org/10.1007/s11948-015-9715-4.

Schwartz, Adina. 'Meaningful Work.' *Ethics* 92, no. 4 (July 1982): 634–46. https://doi.org/10.1086/292380.

Sharkey, Noel. 'Killing Made Easy.' In *Robot Ethics : The Ethical and Social Implications of Robotics*, edited by Patrick Lin, Keith Abney, and George A. Bekey, 111–28. Cambridge, MA: The MIT Press, 2012.

Sharon, Tamar. 'The Googlization of Health Research: From Disruptive Innovation to Disruptive Ethics.' *Personalized Medicine* 13, no. 6 (November 2016): 563–74. https://doi.org/10.2217/pme-2016-0057.

Sparrow, Robert James. 'War without Virtue?' In *Killing by Remote Control: The Ethics of Unmanned Military*, edited by Bradley Jay Strawser, 84–105. Oxford: Oxford University Press, 2013.

Terkel, Studs. *Working : People Talk about What They Do All Day and How They Feel about What They Do*. New York: The New Press, 1972.

Walzer, Michael. 'Liberalism and the Art of Separation.' *Political Theory* 12, no. 3 (1984): 315–30.

———. *Spheres of Justice: A Defense of Pluralism and Equality*. New York: Basic Books, 1983.

Warren, D. Michael. 'Using Indigenous Knowledge in Agricultural Development.' *World Bank Discussion Papers* 127 (1991). https://ideas.repec.org/p/fth/wobadi/127.html.

Wildschut, Diana. 'The Need for Citizen Science in the Transition to a Sustainable Peer-to-Peer-Society.' *Futures* 91 (1 August 2017): 46–52. https://doi.org/10.1016/j.futures.2016.11.010.

Part 3

Control and responsibility

7 Four responsibility gaps with AI

7.1 Who killed Elaine Herzberg?

Six years ago, in Tempe, Arizona, a woman called Elaine Herzberg was struck and killed by a car as she walked her bicycle across the road. The car was an Uber test vehicle in self-driving mode. In other words, at the moment of the incident, the vehicle was driven by a computer, as part of Uber experimentation with automated driving systems on the public road, in compliance with the regulation of Arizona. The technical system did not brake in time and the operator did not intervene to prevent the crash.[1] The legal trial found only one person liable for the killing: Rafaela Vasquez, the vehicle operator employed by Uber. She was not paying sufficient attention on the road, either watching a TV show on her own phone (according to the prosecutors) or monitoring Uber's Slack channel on the handset (according to the defendant). By pleading guilty to endangerment (recklessly putting someone's life at risk), Vasquez was sentenced to three years of supervised probation, with no time in prison. Had she been found guilty of the more serious felony of negligent homicide, as in the original charge, she would have faced four to eight years in prison.[2] From a legal point of view, only one person was found culpable for what seemed to be an avoidable killing.

Moral and political judgements may be harsher though. In fact, authoritative organisations such as the US National Transport and Safety Board explicitly mentioned the lack of a sufficient 'culture of safety' in Uber as a key cause of the incident, de facto attributing some responsibility to them. Arguably, the Arizona regulator also bears some responsibility for the incident with their lax regulations on technology tests in public spaces. A further interesting, related question would be: who is responsible to avoid similar incidents *in the future*? And do we have the conceptual, ethical, and legal resources to answer these questions? In the absence of such resources, we may face the problem of a 'responsibility gap' with artificial intelligence (AI).

DOI: 10.4324/9781003303244-10

The notion of 'responsibility gap' with artificial intelligence (henceforth AI) was originally introduced in the philosophical debate to indicate the concern that 'learning automata' may make it more difficult or impossible to attribute moral culpability to persons for untoward events. This chapter proposes a broader and more comprehensive analysis of the responsibility gap. The responsibility gap, it is argued, is not one problem but a set of at least four interconnected problems – gaps in culpability, moral and public accountability, and active responsibility – caused by different sources, some technical, other organisational, legal, ethical, and societal. Responsibility gaps may also happen with non-learning systems. The chapter clarifies which aspect of AI may cause which gap in which form of responsibility, and why each of these gaps matter. It also proposes a critical review of partial and non-satisfactory attempts to address the responsibility gap: those which present it as a new and intractable problem ('fatalism'), those which dismiss it as a false problem ('deflationism'), and those which reduce it to only one of its dimensions or sources and/or present it as a problem that can be solved by simply introducing new technical and/or legal tools ('solutionism').

7.2 Four responsibility gaps with AI[3]

In 2004, philosopher of technology Andreas Matthias introduced what he called the problem of a 'responsibility gap' with 'learning automata.'[4] In a nutshell, intelligent systems equipped with the ability to learn from the interaction with other agents and the environment will make human control over and prediction of their behaviour very difficult if not impossible. However, human responsibility requires knowledge and control. Therefore, we as humanity are facing a dilemma: either we go on with the design and use of learning systems, thereby giving up on the possibility of human persons being responsible for their behaviour, or we preserve human responsibility, and thereby give up on the introduction of learning systems in society. Many years before, the philosopher Daniel Dennett had titled one of his articles *When HAL Kills, Who's to Blame?*,[5] referring to the iconic episode of Kubrick's *2001: A Space Odyssey* in which the super-intelligent computer HAL 9000 on board a spaceship resists being shut down and tries to kill the spaceship's crew to continue the mission alone. However, now we know we do not need to look at sci-fi for an illustration of the problem: what if an armed drone, after being programmed and activated, could select and engage targets without further human intervention, and civilians are mistakenly killed in an attack? Who would be to blame for that? What if a patient dies following a mistake of a surgical robot with autonomous capabilities? What about the Uber car killing of the opening? Matthias' formulation of the responsibility gap has been quite influential, especially in relation to the development of autonomous weapon systems.[6,7]

More recently, concerns about responsibility gaps were raised more generally in relation to AI. Risks of gaps have been identified around not only the learning capacities of AI but, first and foremost, the opacity, complexity, and unpredictability that these systems generally display.[8] In fact, the question as to what extent persons can or should maintain responsibility for the behaviour of AI has become one of, if not, the most discussed question in the ethics of AI.[9]

However, whereas 'responsibility' is known in philosophy and law for being an ambiguous and polysemantic term,[10] this complexity is rarely reflected in debates on responsibility for the behaviour of AI systems. Therefore, discussions about responsibility or accountability gaps are sometimes partial as they usually appeal to an insufficiently specified notion of responsibility. What do we precisely mean by 'responsibility'? Moral culpability, legal liability, public accountability, or an obligation to prevent future accidents? Moreover, the focus on learning automata or systems equipped with autonomous capabilities may be too limited. Responsibility gaps are due to a multiplicity of factors and sometimes only aggravated by the presence of machines that learn and act on their own. In fact, sufficiently interconnected sociotechnical systems, with limited AI and capacity to learn, but relying on a complex texture of human agents and technical systems, such as bureaucracies or corporates, might also generate responsibility gaps. To understand and address the responsibility gaps problem with AI, it is important to identify its different causes beyond a generic reference to the systems' capacity for 'autonomy,' 'intelligence,' or 'learning.'

The first goal of this chapter is thus to reframe the responsibility gap discussion in terms that are better aligned with the categorisation of responsibility concepts in moral and legal philosophy. By doing so we will be able to better see *what kind of responsibility* is threatened *by which aspect of automation* and *why this matters*.[11] I take some concepts and distinctions from philosophical, legal, and sociological theories of responsibility [12] and use them to identify four different kinds of responsibility gaps: the *culpability* gap, the *moral accountability* gap, the *public accountability* gap, and the *active responsibility* gap. I will also identify, for each of those, different possible causes, integrating Matthias' classic analysis that identified learning capacities of automata as the main source of responsibility gaps, generically speaking. For a deeper look, Matthias addressed in his work what I will call the culpability gap – the risk that no human agent might be legitimately blamed or held culpable for the unwanted outcomes of actions mediated by AI systems. Gaps in this kind of responsibility have already received some attention, both from a moral[13] and a legal perspective.[14] Attention also went to the accountability gap in relation to autonomous weapon systems[15] and more generally within the discussion on the explainability of algorithms and AI.[16] However, I propose to distinguish two forms of the accountability gap: the public accountability gap, i.e. citizens not being able to get an explanation for decisions taken by public agencies, and a broader moral accountability gap – i.e. the reduction of human agents' capacity to make sense of and explain to each other the 'logic' of their behaviour. This can be due to the mediation of opaque, unexplainable algorithms and complex systems with autonomous capabilities as well as the lack of appropriate psychological, social incentives, or institutional spaces that promote these explanations. One particularly important form of the moral accountability gap concerns the difficulty for engineers and other agents involved in the process of technological development to systematically discuss their understanding of the goal and meaning of this process with each other. Finally, there is what I call the active responsibility gaps. These gaps arise when persons designing, using, and interacting with AI are not sufficiently aware, capable, or motivated to see and act according to their moral obligations towards the behaviour of the systems they design, control, or use. In particular, they fail to ensure that these systems do not impact negatively on the rights and interests of other persons but, ideally, contribute positively to their well-being. Distinguishing four responsibility gaps and their various sources is the focus of the first part of this chapter. Table 7.1 summarises the different senses of responsibility and their value.

In the second part, I will discuss some of the common approaches that have so far been taken towards addressing responsibility gaps. Those approaches, as I will argue, offer a partial and limited understanding of the responsibility gap, and thus offer solutions that would apply only to specific aspects of them. The so-called 'fatalist' approaches[17] tend to focus on a too limited understanding of the responsibility gap, which is a gap in *culpability* for the behaviour of *learning* technological automata. The so-called 'deflationist' approaches[18] underestimate the novelty of AI systems and their implication for culpability attributions in morality (blameworthiness) and the law (liability); they also seem to underestimate the risks of gaps in the moral and political accountability of system designers as well as gaps in theirs and other agents' active responsibility for the behaviour of artificial intelligence. Promoters of 'explainable AI' and other scientific and technological improvements tend to ignore the psychological, social, and political dimensions of the interaction with AI systems, thereby running the risk of embracing some form of 'technical solutionism,'[19] by which all the moral and social problems of human responsibility for the behaviour of AI systems can be fixed simply by improving the working of AI techniques. Lawyers and policymakers proposing the revision of current legal liability regimes (including the extension of strict and product liability regimes and 'electronic personhood') may either underestimate the importance of maintaining some form of human *moral* responsibility for the behaviour of the AI or recognise this need but without saying how moral and social practices – and not only legal rules – should change in order to govern a responsible transition to the use of AI. I call this the risk of 'legal solutionism.' One result of this critical review is the recognition that the different notions of responsibility, though distinct, are also interconnected and that addressing one kind of gap often requires paying attention to one or more of the others. This suggests the necessity of a more integrated and comprehensive approach.

7.3 Varieties of responsibility gaps

The term 'responsibility' has different meanings. Legal philosopher Herbert Hart's classical account[20] lists four senses of responsibility: role-responsibility, causal responsibility, capacity-responsibility, and liability-responsibility. Based on more recent work in moral philosophy by John Gardner,[21] by Mark Bovens,[22] and by Ibo van de Poel et al.,[23] I propose a revised and modified list of four forms of responsibility that are particularly relevant – but not limited to – the context of automation and AI. The four senses of responsibility I will work with are: culpability, moral accountability, public accountability, and active responsibility. The next four sub-sections present these four forms of responsibility and identify the corresponding challenges created or furthered by the introduction of AI (Table 7.1). Some examples will serve as illustrations.

7.3.1 Culpability gaps

When things go wrong and important interests or rights such as the right to physical integrity or life are infringed, we, as victims and as society, not only want to understand what happened but also why it happened. We want to know whether the harm was the result of someone's wrongful behaviour, and, if it turns out that there is no justification or excuses for this behaviour, we want the guilty party to be condemned, or even punished for their behaviour. This is, in a nutshell, culpability or blameworthiness. Whether and to what extent attributing each other culpability makes sense at all has been the main subject of the centuries-long philosophical debate on free will, typically in the light of causal determinism (is it fair to blame each other if all our actions are necessarily caused by previous physical and mental events?) or more generally in the light of views about human behaviour that are shaped by (neuro)scientific knowledge (if human action can be fully explained by behavioural/social/neuroscience, what is left of moral culpability?).[24] However, many philosophers and most lawyers and laypersons do believe that (fair) social and legal practices of attribution of culpability should be maintained and promoted, in addition to forms of social and psychological education or therapy,[25] at least insofar they are

Table 7.1 Types of responsibilities and gaps in the context of AI

Type of responsibility	*Definition*	*Gaps with AI*	*Example*
Culpability	Blameworthiness for wrongdoing based on intention, knowledge, or control	AI-making prediction and control more difficult, thereby creating new legitimate reasons/ excuses for wrongdoing	An avoidable road crash involving an automated driving system that nobody could individually predict or prevent
Moral accountability	Duty of humans to explain one's reasons and actions to others (under some circumstances)	AI-making processes unexplainable to the very persons using it	A doctor not being able to explain the reasons for their AI-supported diagnosis to a patient
Public accountability	Duty of public agents to explain their actions to a public forum	AI shifting discretionary powers towards IT experts and data analysts (often outsourced to private companies), whose work is harder to publicly scrutinise	A government using (private) AI systems in support of their decision-making
Active responsibility	Duty to promote and achieve certain societally shared goals and values	Actors involved in the design or use of AI not being sufficiently aware of their own responsibility to prevent harm deriving from AI or not being able or motivated to fulfil this obligation	Engineers or managers only looking at the technical benefits of AI

the legitimate expression of appropriate moral sentiments by the wronged individuals and society at large, [26] they reinforce the social commitments to shared norms[27] and, possibly most importantly, they contribute to control and reduce undesirable behaviour. Similarly, State-administered punishment for serious criminal behaviour may be morally defensible and even desirable, insofar as it gives effectiveness to the expression of public condemnation[28] and serves the utilitarian goals of discouraging similar behaviour in the future by the defendant themselves and in general by other people (the so-called deterrence function of punishment). Finally, (public) attributions of culpability have the function to compensate the victims, symbolically or materially, typically in the case of compensations to plaintiffs in civil litigations.

AI systems may put culpability practices under stress in different ways, preventing the achievement of one or more of the goals of these practices. Consider, as an example, self-driving cars like that of the example opening this chapter. First, self-driving cars may make the network of agents involved in the process of driving more complex, just because more agents are involved and/or new forms of interactions are created. For instance, a vehicle may be operated by driver D1, with the assistance of the automated driving system AS, produced by the car manufacturer X, powered with digital systems developed by the company Y, possibly including some form of machine learning developed by the company Z, and enriched by data coming from different sources, including the driving experience of drivers D2, D3 … D*n*; vehicles in this system are in principle subject to a standardisation process done by the agency S, the traffic is regulated by the governmental agency G, drivers are trained and licensed by the agency L, etc. Second, some specific features of present-day learning AI systems may make this interaction particularly unpredictable – typically when the vehicles' performance is potentially re-designed by the second on the basis of new data acquisition and processing – and opaque, if the reasoning scheme underlying systems' actions is not easily accessible to their controllers, regulators, or even their designers.

Agents operating in such a socio-technical system (designers, programmers, drivers, regulators, bystanders, etc.) may more easily find themselves acting wrongly, for instance, causing an avoidable road crash while at the same time having a legitimate excuse. Nobody, and certainly not they, could reasonably predict certain circumstances or reasonably avoid a certain outcome, therefore not being open to legitimate blame.[29] I call this the 'culpability gap.'

The culpability gap has not been created by the introduction of 'learning automata' (machine learning) and their inherent unpredictability, as it has been framed, for instance, by Andreas Matthias. As a matter of fact, other intelligent, autonomous entities with 'no soul to blame and no body to kick,'[30] such as, e.g., bureaucracies and corporates, may in themselves generate gaps in culpability.[31] This has been classically identified as 'the problem of many hands.'[32] AI systems play, however, an important role by contributing to new versions of the phenomenon and thus making it more visible. Also, the use of AI and data-driven machine learning in decision-making importantly introduces a new element of technical opacity and lack of explainability that makes it more difficult for individual persons to satisfy the traditional conditions for moral and legal culpability: intention, foreseeability, and control.[33]

Culpability gaps are problematic insofar as the more persons designing, regulating, and operating the system can legitimately and possibly systematically avoid blame for their wrong behaviour, the less these agents will arguably be incentivised to prevent these wrong behaviours by striving for a high(er) level of safety, awareness, attention, motivation, and skilfulness. And the less victims of unjust harm will be likely to receive compensation. Also, it might become more difficult for persons to make sense of their moral sentiments in relation to wrongs and accidents and to direct their reactive attitudes towards some legitimate target. This may impoverish the human capacity to express moral judgement and may feed helplessness and moral scepticism towards the possibility of understanding and rectifying wrongdoing. As noted by John Danaher, the desire to find a scapegoat to satisfy these feelings may also be fuelled.[34]

7.3.2 Moral accountability gaps

Culpability is a particularly serious form of (moral) responsibility, but it's not the only one. Individuals are often called upon to explain their choices or actions in less threatening ways, for instance, when family, friends, or acquaintances ask them 'why' questions, not necessarily with the intention of judging or blaming them, but possibly to just engage in a conversation and to better understand their reasons and expectations. Why were you late at the appointment, why did you start taking guitar classes, and why did you turn down that job offer? I call this expectation to answer (at least some) 'why' questions by other persons 'moral accountability,' to distinguish it from the 'public accountability' discussed below. Moral accountability has been presented in the philosophical literature on moral responsibility as a key aspect for the justification and understanding of moral responsibility practices.[35] The legal philosopher John Gardner calls it 'basic responsibility' because he sees it as the core of what it means to be a rational and reflective person in society.[36] In this sense, being accountable, unlike being culpable, is something to be desired rather than avoided as it is a constitutive part of being able to reflect on one's actions and to participate in meaningful social relations. It also helps people to see events in the world as being connected to their own rational capacities, thereby supporting their sense of agency and responsibility.[37] It is a classic view of human responsibility, which can be followed back to the old Socratic motto 'know thyself.' In fact, Gardner calls this the 'Aristotelian story' about responsibility, as it focuses on people's capacity to make sense of their and others' actions and choices as something connected to their abstract reasons (as opposed, for instance, to just physical or biological events).[38]

Moral accountability also has an instrumental value. The process of exchanging questions and answers helps finding explanations for things that have happened and reinforces trust and social connections between individuals. The mere prospect of potential requests for explanation and justification also tends to reduce undesired behaviour by pushing persons to be more clearly aware of the impact of their actions on others and therefore motivated to prevent unwanted outcomes (and potential blameworthiness). In relation to engineering practices, Genus and Stirling[39] have recently stressed the importance of Collingridge's proposal to 'engage more strongly with accountability in debates bearing on key elements of responsible innovation.'[40] In their view, also inspired by Lindblom,[41] accountability is a key tool to enhance the reflexivity of the agents involved in the design and development of (new) technologies and to promote responsiveness between them and those who will be affected by their creations. In the literature on 'responsible research and innovation' the importance of accountability practices has been recently emphasised and categorised under the label of 'responsiveness between stakeholders.'[42]

Moral accountability may be blurred in different ways by the introduction of AI. First, in general and similar to what was observed above about culpability, by contributing to create a more complex chain of decision-making and actions, AI may make it more difficult for individual agents to make sense of the reasons why a certain decision was taken, what their exact role in the operation was, and, in general, whose reasons and what reasoning were governing the system they were part of. However, data-driven machine (deep) learning, due to its intrinsic opacity, might make a system's behaviour extra hard to understand and explain. An example of AI affecting moral accountability specifically due to the opacity and complexity of AI may be that of a medical doctor using an AI-driven system for diagnosing. These systems are usually based on deep learning techniques that require a thorough training over a dataset, the nature of which is well known and clear. In other words, the system will train on a set of well-known, well-established cases, before being applied to new and unknown cases. The way knowledge is represented in the machine and the exact way the machine distinguishes between a positive and negative diagnosis are not only inaccessible to the doctor who uses the system and to an important extent also to those who designed it.[43] Therefore, the capacity of the different agents, including the users, to make sense of the 'logic' of the system's behaviour may be weakened and sometimes lost, together with their capacity and willingness to engage in a meaningful conversation about their role and the responsibility that comes with it.

In addition, the whole process of technology development and production is arguably pervaded by an increasing pressure towards deploying proprietary technologies that, even when working through mechanisms accessible to their developers and programmers, are purposefully designed to be inaccessible to public scrutiny and the users themselves.[44] Technology developers, driven by the desire to both minimise industrial espionage and maximise customer loyalty (by, e.g. binding them to their assistance network), will usually avoid sharing data and engineering insights. This is one important element of what Frank Pasquale, playing with the ambiguity of the phrase, calls the 'Black Box Society.'[45] Our society filled with digital intelligent services and systems is a black box in the sense of a recording machine like those present on aeroplanes, from the point of view of the companies who own and control the data-driven AI systems, and a black box in the sense of an impenetrable object for the people who use these systems.

The general concern is that AI may make individual persons less able to understand, explain, and reflect upon their own and other agents' behaviour. Let's call this the 'moral accountability gap'.

7.3.3 Public accountability gaps

One specific form of accountability is attached to politicians, civil servants, and other agents with a public function: public accountability. Public accountability, in public administration scholar Mark Bovens' classic definition, is a 'relationship between an actor and a forum, in which the actor has an obligation to explain and to justify his or her conduct, the forum can pose questions and pass judgement, and the actor may face consequences.'[46] According to Bovens, effective mechanisms of accountability may enhance both the effectiveness of a complex public decision-making system and its compliance with liberal-democratic values.[47] Accountability promotes democratic control (transparency), limits abuses of power (corruption), and also brings more effectiveness to the institutions. In fact, providing administrators with information about their own functioning and forcing them to reflect on their successes and failures will eventually allow and encourage them and others to improve their future performances.

There has recently been a big legal and political debate on the extent to which the introduction of AI-based automated (administrative) decision-making is desirable and legitimate.[48] Also, it has been called into question if the EU's General Data Protection Regulation (GDPR) provisions guarantee sufficient transparency and accessibility to these mechanisms for those who are subject to them.[49] Most of these discussions point to the fact that algorithmic decision-making is often difficult to explain to human agents, due to the different and sometimes inscrutable ways of AI operations: the so-called 'black box' problem.[50] However, Noto La Diega correctly notices that issues of explainability may arise not only due to technical black boxes but also due to what he calls organisational and legal black boxes created or aggravated by the introduction of AI in public administration.[51]

Scholars in public administrations have further explained the sources of these organisational and legal black boxes and their relevance to public accountability.[52] Traditionally, public agencies were organised as 'street-level' bureaucracies. Processes were managed by individual case managers who had direct contact with individual citizens and substantial discretionary powers.[53] With the introduction of digital decision-making systems, these discretionary powers of street-level professionals have been limited. However, this has also greatly shifted the locus of administrative discretion from individual public officers to IT experts, responsible for programming the decision-making process and translating the legislation into software, and to the data analysts, responsible for the acquisition and analysis of data.[54] Moreover, these 'system-level' bureaucracies are part of larger networks and chains of delegation in which data are exchanged and reused.[55] This shift raises three challenges for public accountability. First, the development in information technologies is often outsourced to private parties or tech giants such as Google, who are not publicly accountable and may not be willing to disclose critical information about the functioning of their systems.[56] For example, in her book *Automating Inequality*, Virginia Eubanks[57] reports stories of private contractors not being able or willing to disclose the reasons for the failures or mistakes in their digital systems used by some US States for welfare benefit allocation procedures. Second, more generally the work of the software engineers and data professionals in public organisations is usually not visible and subject to public and legal scrutiny. Finally, far more data are exchanged between many different (public) organisations than in the past. In this way, the introduction of AI makes the 'problem of many hands'[58] more acute: data coming from different sources are introduced and enriched at different points in the data chain. Individual citizens may have a hard time finding out who they should turn to, if data are incorrect, corrupted, or biased as a collective outcome of a series of minor contributions. Technical, organisational, and legal black boxes are the sources of what I call the 'public accountability gap'.

7.3.4 Active responsibility gaps

The philosophy literature on the professional responsibility of engineers usually distinguishes between 'active' and 'passive' responsibility.[59] In a nutshell, active responsibility is forward-looking and concerns the goals, values, and (legal) norms that professionals such as engineers are supposed to promote and comply with as well as the consequences they need to prevent and avoid. Passive responsibility is backward-looking and concerns the moral and legal consequences engineers must face in case something goes wrong. The three forms of moral responsibility discussed above – culpability, moral accountability, and public accountability – are all forms of passive responsibility. The legal duty to provide high standards of safety and the so-called corporate social responsibility of companies are typical examples of active responsibility.

One well-known problem with the active responsibility of engineers is that while they arguably have an individual active responsibility to promote societal goods, their work is most often embedded in networks of different agents and institutions.[60] They may, for instance, be involved in projects connecting scientists, academic institutions, and technology companies operating in the market. As seen above, this may create problems for the attribution of passive responsibility, as in case something goes wrong it may be easy (and sometimes legitimate) for individual engineers to shift responsibility to other agents or institutions in their network. Sometimes it may be even the case that nobody can legitimately be held responsible for a specific unwanted event.[61] What is often overlooked is that the networked nature of the engineering work may also create issues with the attribution of active responsibility. As noted by the philosopher of technology Udo Pesch, engineers may not have a clear and consistent representation of what their (social) role is: are they scientists, technicians, or business persons? They may be unclear about what goals and values they should strive for: truth, innovation, market shares? They may not even have clear and shared systems of principles, norms, and rules to follow in their profession, or they may lack sufficient capacity or motivation to reflect upon and interpret these rules in concrete cases.[62] What we call 'the active responsibility gap' of engineers has thus old historical roots. Institutions like the modern State, the market, and science have since the time of their birth been the object of systematic intellectual reflection by statesmen, scientists, philosophers, and others, and they were therefore seen and built according to a complex 'intellectual template.' Modern technology development, on the other hand, 'has never been integrated into this template.'[63] In fact, whereas political philosophy, philosophy of science, and political economy are well-established disciplines with deep historical roots, philosophy of engineering is in comparison a relatively new and not even fully established and recognised academic discipline. This problem concerns primarily engineers but, as we will see in more detail below, also other professionals involved in the design, development, regulation, control, and even use of technology.

Based on the general framework above, I propose that the introduction of AI may create two different but related sets of issues. First, engineers and other agents involved in the development and use of technology may not be (fully) *aware* of their respective moral and social obligations towards other agents. Think, as an example, of a manager of an IT company who, as a result of their personal education or the engineering and business culture in which they have been raised, is genuinely and sincerely convinced that they are benefitting the public by providing them with more comfort through the use of their new products. And, in any case, it is not their responsibility to try to minimise the possible negative impact of the use of these products on people's well-being, privacy, or political freedom.[64] In van de Poel and Sand's classification of active responsibility,[65] this is a gap in 'obligation.' Second, engineers and other agents involved in the development and use of technology may be well aware of an obligation yet not *sufficiently able or motivated* to fulfil it. Think, as an example, of military personnel using a new AI-based weapon system: while being perfectly aware of their general obligation to use the system in compliance with the requirement of international law, they may end up making an illegal use of the system, due to insufficient technical training and/or insufficient capacity to resist pressures to use the technology in a certain way, coming from their superiors or others around them. In van de Poel and Sand's classification, this is a gap in 'virtue,' i.e. the concrete capacity and inclination to perform according to certain norms and principles. Let's call these two issues, taken together, the 'active responsibility gap.'

7.4 Partial answers to responsibility gaps: fatalism, deflationism, and solutionism

The previous section discussed how considering different senses of responsibility allows one to highlight the existence of four different kinds of responsibility gaps with AI, summarised in Table 7.1. The problem of responsibility gap with AI, as it turns out, is not one problem but a set of at least four interconnected problems – gaps in culpability, moral accountability, public accountability, and active responsibility. Moreover, these gaps are caused by different sources, which can be old (i.e. the complexity of social and technical systems), new (i.e. the data-driven learning features of present-day AI), more technical (i.e. the intrinsic opacity of algorithmic decision-making), more political and economic (i.e. the implicit privatisation of public administration), or more societal (i.e. the engineers' and other actors' lack of awareness of or capacity to comply with their (new) moral, legal and societal obligations). Sufficient awareness of this complexity has been missing in the debate so far.

Current debates have put strong emphasis, from time to time, on just one of these different problems and on one or two of their sources, thereby often not giving sufficient attention to the broader picture. In this second part of the chapter, I make a critical revision of a representative sample of the literature on the responsibility gap, with two goals. First, to show the extent to which this literature misses out on the complexity of the responsibility gap. Second, to suggest that taking such a partial approach to the responsibility gap risks offering not only an incomplete picture of the problem but also a distorted one. I present three possible distortions in the presentation of the responsibility gap with AI, which I call, respectively: 'fatalism,' i.e. the idea that the responsibility gap is a new and intractable problem, 'deflationism,' i.e. the idea that the responsibility gap is not new and not a problem, and 'solutionism,' i.e. the idea that the responsibility gap is a problem that can be solved by simply introducing new technical and/or legal tools. To be clear, I am not claiming that every author discussing just one aspect of the responsibility gap necessarily falls victim to one of these distortions: as a matter of fact, many authors are well aware that they are addressing only one aspect of a more complex problem. What I am suggesting is that, in the long run, focusing only on these partial analyses may provide a misleading picture of the responsibility gap, thus hindering the creation of an appropriate response.

7.4.1 Fatalism and deflationism

At the two extremes of the debate on responsibility gaps are those whom I call, respectively, fatalists and deflationists. The fatalist approach is best captured by Andreas Matthias' 2004 paper on 'The Responsibility Gap.'[66] According to Matthias, the introduction of 'learning automata' in society poses an unprecedented challenge to moral responsibility (i.e. culpability) and presents us with a moral dilemma. Culpability requires knowledge; learning systems make knowledge (i.e. prediction of outcomes) impossible, therefore learning automata make it impossible to (legitimately) attribute culpability to human persons for the actions mediated by learning systems. We as a society are then facing a dilemma: either we introduce learning systems and give up on culpability, or we maintain culpability, and give up on the introduction of learning systems in society.[67] Matthias' formulation of the responsibility gap has been quite influential, especially in relation to the development of autonomous weapon systems.[68] In this perspective, the responsibility gap created by learning automata (AI) is a new, serious, and intractable problem. I have argued above that the culpability gap is not completely new, and I will suggest in the next chapter how it could be potentially addressed by designing sociotechnical systems with a new notion of human control ('Meaningful Human Control as reason-responsiveness').

At the other extreme of the debate are those who believe that the culpability gap for learning automata and AI more generally is not a problem after all, and anyway not a new one, so that old technical, moral, and legal recipes will just suffice. I call these 'deflationists.' Some deflationists just embrace the first horn of Matthias' dilemma. If we have reasons to believe that the introduction of learning automata will bring significant societal benefits, for instance, in terms of efficiency, effectiveness, or overall safety of the processes, we may and should introduce these in society, even if this will lead to some erosion of human moral responsibility (in any possible sense), for the – *ex hypothesi* – reduction of the number of accidents. For instance, many believe that if the introduction of AI in critical tasks such as medicine, transport, and warfare is likely to reduce the overall number of deaths or injuries by reducing the impact of human error, then we should not care too much about the risk of gaps in moral responsibility. This is a too simplistic utilitarian approach, which tends to obfuscate the relevance of individual moral and legal duties and rights. Individual people have specific duties to minimise *their own* contribution to wrongdoing, and specific rights to be individually protected from wrong harms. These duties and rights must arguably be assessed independently from the calculus of the overall net increase or decrease of societal welfare.[69] That approach also ignores the moral relevance of fairness and distributive justice in assessing the moral risks of technology, by asking how the benefits and risks are distributed, who is benefitting from and who is harmed by new technologies.[70]

However, philosophers Simpson and Müller[71] have proposed a more sophisticated version of deflationism, one which embraces the first horn of Matthias' dilemma (accepting learning systems *and* some gaps in moral culpability), while trying to pay due respect to individual rights and distributive justice. The argument goes like this. AI may create some culpability gaps, but this is not new. Non-intelligent and non-learning systems like bridges and buildings can also collapse (fail) without any human person being culpable for that, and we as a society find this acceptable, insofar as all reasonable precautions had been taken, and nobody could have reasonably prevented the accident. The same can be said about AI. In contrast to simplistic utilitarian approaches, this approach claims that we as a society have two obligations: to reduce the aggregate risks involved in the use of technology and to minimise the risk of harm to *each* of the persons involved. If these two goals are achieved by the introduction of a given technology, and a 'tolerance level' for failure is set that is 'as low as technologically feasible,'[72] then it will not be a moral problem if accidents happen for which no one is culpable (i.e. there will indeed be a culpability gap), because all relevant considerations in terms of safety, justice, and responsibility have been respected.

I agree that culpability should not be pursued always and at all costs and that some culpability gaps are unavoidable and morally acceptable. However, I doubt that deflationists have yet identified a fair system to decide which culpability gaps with AI systems are morally acceptable. Indeed, justice and rights can be preserved by setting reasonable standards of care, which may also allow for fair culpability attribution and the prevention of unwanted responsibility gaps. But this is a statement of the problem, not a solution: it is close to begging the question. To what extent it is possible to establish standards of reasonable care for the design, use, and regulation of AI in the same way as for buildings and bridges is precisely the question raised in the ongoing (legal) debate on the responsibility gaps for AI.

In relation to the US system, lawyer Ryan Calo pointed out that culpability gaps with AI may happen precisely because the traditional assumptions about what should count as sufficient intention, knowledge, and foreseeability on the side of the defendant (in criminal law) may not apply, due to the emergent and unpredictable behaviour of AI.[73] Also, traditional assumptions on designers or managers having 'exclusive control' of artefacts can hardly apply in relation to systems whose behaviour is influenced by a chain of different actors (manufacturers, programmers, and users).[74] How to achieve a fair and right-respecting distribution of risk and culpability in this new context is still open to discussion. Moreover, the *accountability* gap potentially created by the unpredictable, opaque, and interactive behaviour of AI systems may also make it difficult to establish a priori whether the system will be able to comply with the requirement of fairness: designers and users of AI may simply not be in the position to know that 'their' system is discriminatory or otherwise unfair in the first place.

A related concern with deflationist proposals is that by stating that the 'threshold of safety' eventually determining the culpability attribution should be as low as technologically feasible, they risk encouraging the belief that this is mainly a technical question, one that is up to engineers and experts to solve. However, giving engineers and other experts the power to decide on this threshold may result in a technocratic approach, sharpening rather than solving the culpability gap: technical experts may (honestly) believe that nobody is to blame for an accident because they have done what could reasonably be expected given the current capabilities of the technology, but this may not match with the well-informed judgement of a non-expert and the moral and legal requirements of society. In fact, recent legal history of the standards of negligence in medical practice shows a trend towards abandoning a system in which courts rely exclusively on expert professional opinion for their assessment of professional malpractice. This is due to the recognition that professional opinion may sometimes be unreasonable or irresponsible, too conservative, biased, or otherwise reflecting the interests of the members of the profession rather than the interest of the public. Also from a normative point of view, courts being dictated by experts was considered a dangerous shift towards technocracy and not in line with a right-based society.[75]

Relatedly, from a broader perspective, the old-style division of roles – engineers give facts to regulators and regulators establish whether the technology is safe enough – does not incentivise mechanisms of moral accountability between engineers and societal stakeholders,[76] as they may not sufficiently promote a well-informed public deliberation on what a 'reasonable threshold' of safety (and other values) in emerging technology should be. Nor does this approach incentivise engineers to go beyond the current state of the art in technology and look for innovative solutions which may improve the capabilities of current technology to better satisfy complex and potentially conflicting societal demands[77] such as higher levels of safety combined with better predictability. In the terminology introduced above, deflationist strategies fail to address gaps in the active responsibility of engineers. They may also fail to address gaps in (public) accountability if they accept delegating the setting of a reasonable standard of care to experts alone.

7.4.2 *The risks of solutionism*

Others recognise the novelty of the responsibility gap but do not believe in its intractability, trying to offer some new technical and legal solutions to address it. Whereas some of these solutions might in principle be part of a comprehensive strategy to address the problem of responsibility gap in its entirety and complexity, their authors often fall short of providing a comprehensive plan. Moreover, even when this is not the intention of their authors, these proposals even run the risk of fuelling the temptation of 'solutionism,'[78] which is the belief that all complex socio-technical and political problems can be 'solved' (or avoided) by the introduction of new techniques.

By distinguishing two main approaches to address the responsibility gaps, i.e. the technical and the legal, I suggest that if taken in abstraction from the broader picture presented above, these run the risk of leading, respectively, to 'technical solutionism' and 'legal solutionism.' Below I discuss both types of solutionism.

7.4.3 *Explainable AI and technical solutionism*

One of the commonly recognised causes of generically defined responsibility gaps is, as discussed in the previous sections, the lack of transparency, explainability, and interpretability of machine-aided decision-making, be it defined as just algorithmic or properly AI. Although there being multiple senses and extents to which a system can be said to be explainable,[79]

I will stick to the most basic form, where 'a user cannot only see, but also study and understand how inputs are mathematically mapped to outputs,'[80] and where transparency of the whole process is granted. Theorists identified transparency and explainability of algorithms and AI as important elements to safeguard the 'traceability' of human responsible agents and, consequently, a fair attribution of moral responsibility.[81] I believe that algorithmic explainability, though constituting one interesting element in a complex strategy to address the responsibility gaps, is neither a sufficient nor a necessary condition to address them. Believing the contrary would amount to what I called 'technical solutionism': the belief that new technological solutions may be sufficient in themselves to address complex socio-technical and political problems.

One reason why explainability is not sufficient to address the responsibility gaps is that, as discussed, such gaps are due to different problems that are not entirely addressed by increasing transparency and explainability of the algorithmic parts of a system. Mittelstadt and colleagues,[82] echoing Judith Simon,[83] correctly pointed out how an important factor determining traceability, still under-researched, is the fact that responsibility is distributed or diffused 'across a network of human and algorithmic actors simultaneously,' which is closely related to what I previously referred to as 'the problem of many hands'[84] and, more generally, 'organisational black boxes.'[85] Relatedly, algorithmic transparency and explainability do not necessarily allow for fair attribution of moral and legal culpability. Very complex systems might, in principle, be understandable and open to scrutiny, but perhaps only after the fact, by a very selected audience, and over a relatively large time frame. But this does not mean that this behaviour may be sufficiently understood and predicted in advance by any of the human agents involved in their design, use, or regulation. Nor does this entail that any of these agents have been given sufficient awareness and capacity to comply with some specific obligations, to prevent some outcome, to explain it once it has happened, or all of these.

Explainability might in other cases also not be necessary to address culpability, accountability, and active responsibility gaps. Some culpability and accountability gaps with AI can potentially be addressed by providing some agents along the chain of design, development, regulation, and use with sufficient knowledge of the limitations of the technical systems (including their opacity, etc.) and a sufficient awareness of their obligation to prevent unwanted results in the deployment of such technologies. That is, by promoting their active responsibility.[86] In the presence of sufficient knowledge and training, for instance, a military commander can be reasonably held accountable and culpable for his conscious decision to deploy an unpredictable technical system in a military mission, which ends up unlawfully killing innocent civilians. Similarly, the manager of a car-manufacturing company and/or the chair of a road safety agency can be legitimately held accountable and culpable for their decision to put/allow on the public road a self-driving vehicle whose behaviour, as they well knew, could not be sufficiently predicted and explained. In a relevant sense, they could and should have known better.

7.4.4 New liability regimes and the risks of legal solutionism

Some legal scholars and policymakers have recently recognised that the introduction of AI systems may potentially increase the number of accidents and/or introduce new kinds of accidents and/or increase the number of accidents for which victims may not receive compensation, due to the difficulty of applying existing legal liability regimes, typically negligence and product liability, to any of the actors involved in the network of design and use of new technologies.[87] This is the legal culpability gap discussed in Section 7.3.1. To address these issues, these scholars and policymakers have committed themselves towards revising or introducing new liability mechanisms, which may allow for compensating victims of accidents involving AI for which no clear human fault can be attributed. The mechanisms include faultless compensation schemes for damages caused by AI systems[88] and the introduction of electronic personhood.[89] However, an exclusive focus on bridging the liability gap may be insufficient and potentially self-defeating from the point of view of the broader plan of bridging the responsibility gaps.

As explained in Sections 7.3.1 and 7.3.4, there are several reasons why we might want to preserve fair practices of attribution of moral blame and culpability as well as 'active responsibility' practices in addition to fair and effective practice of legal liability and compensation schemes. Answering the question 'who should be legally punished' or 'who pays' for wrong AI-mediated decisions and behaviours is not sufficient to answer the broader question 'who is responsible for them and how to prevent these outcomes in the first place.' Liability regimes grounded in individual culpability or fault (such as criminal liability and criminal and civil liability by negligence) may be well suited to deal with clear and bold individual responsibilities. However, they may be less adequate to cope with substantial shared responsibilities derived from manifold individual small faults. According to an example by van de Poel and colleagues, it would probably make no sense to hold individual people liable for their share of pollution, but that does not mean that they cannot be blamed or shamed for it, or that other policy and legislative tools should not be used to discourage individual and corporate behaviour from causing further pollution.[90] The same can be said about the effects of digital technologies.

Faultless liability regimes and legal personhood of artificial agents risk shifting away attention not only from culpability and accountability but also from active responsibility. In fact, those approaches might underestimate the importance of promoting proactive, preventive approaches to create safe and societally beneficial technical systems. Liability regimes are managed by the State and require strict standards of causation, evidence, and seriousness. Effective as they may be in (dis) incentivising some behaviour, those regimes do not and cannot cover all undesirable behaviour. It is not possible or desirable, e.g., to have the State checking and judging professional behaviour, but that does not mean that anything goes. Since risky behaviours without (provable) harm would fail to be sanctioned under a liability scheme, professionals' good conduct can only be granted by relying on their own awareness and knowledge of their moral and legal responsibility towards society, and their individual capacity and motivation to comply with it. Moreover, corporate or civil liability may not be a strong incentive to behave, e.g. for agents or companies who can easily afford to pay any fine or compensation or who may rely on the difficulties to enforce legal norms in this field.[91] Instead, blaming and shaming and strong political initiatives may sometimes be more effective tools.

In addition to adjusting and revising regimes of liability, we also need to create better mechanisms to promote the moral accountability of all agents involved in the design and use of AI systems, better mechanisms of public accountability for those who design and regulate AI systems operating in the public space, and, possibly more importantly, mechanisms and policies to promote a better culture of active responsibility of all the designers, managers, controllers, and users of AI systems.

7.5 The need of a more comprehensive approach to responsibility gaps

The problem of responsibility gap is not one problem but a set of at least four interconnected problems – gaps in culpability, moral and public accountability, and active responsibility – caused by different sources, some technical, other organisational, legal, ethical, and societal. Partial approaches to the responsibility gap – i.e. focusing only on one form of responsibility and/or only one source of gaps – bring the risk of offering not only an incomplete picture of the problem but also a distorted one, one that hinders the creation of an appropriate response. To avoid these risks, a more comprehensive approach to human control and responsibility is needed. In the next chapter, I will introduce the concept of Meaningful Human Control as reason-responsiveness, which I consider to be a very promising approach to address the issue of responsibility gaps with AI as well as the more general concern that AI systems may get out of human control, causing unwanted risks and harm to people.

Notes

1 Jack Stilgoe, 'Who Killed Elaine Herzberg?,' in Jack Stilgoe, *Who's Driving Innovation?* (Cham: Palgrave Macmillan, 2020), 1–6, https://doi.org/10.1007/978-3-030-32320-2_1.

2 Lauren Smiley, 'The Legal Saga of Uber's Fatal Self-Driving Car Crash Is Over,' *Wired*, accessed 16 October 2023, https://www.wired.com/story/ubers-fatal-self-driving-car-crash-saga-over-operator-avoids-prison/.

3 This chapter reproduces in large part the content of Filippo Santoni de Sio and Giulio Mecacci, 'Four Responsibility Gaps with Artificial Intelligence: Why They Matter and How to Address Them,' *Philosophy & Technology* 34, no. 4 (14 December 2021): 1057–84, https://doi.org/10.1007/s13347-021-00450-x. Credit is therefore due to Giulio Mecacci for the content of this chapter.

4 Andreas Matthias, 'The Responsibility Gap: Ascribing Responsibility for the Actions of Learning Automata,' *Ethics and Information Technology* 6, no. 3 (1 September 2004): 175–83, https://doi.org/10.1007/s10676-004-3422-1.

5 Daniel C. Dennett, 'When Hal Kills, Who's to Blame? Computer Ethics,' in *Hal's Legacy: 2001's Computer as Dream and Reality*, ed. David G. Stork (Cambridge, MA: The MIT Press, 1997), 351–65.

6 Robert Sparrow, 'Killer Robots,' *Journal of Applied Philosophy* 24, no. 1 (2007): 62–77, https://doi.org/10.1111/j.1468-5930.2007.00346.x.

7 Human Right Watch, 'Mind the Gap: The Lack of Accountability for Killer Robots,' 2015, https://www.hrw.org/report/2015/04/09/mind-gap/lack-accountability-killer-robots.

8 Brent Daniel Mittelstadt et al., 'The Ethics of Algorithms: Mapping the Debate,' *Big Data & Society* 3, no. 2 (1 December 2016): 205395171667967, https://doi.org/10.1177/2053951716679679.

9 Sven Nyholm, 'Attributing Agency to Automated Systems: Reflections on Human–Robot Collaborations and Responsibility-Loci,' *Science and Engineering Ethics* 24, no. 4 (18 August 2018): 1201–19, https://doi.org/10.1007/s11948-017-9943-x; Matthias Braun et al., 'Primer on an Ethics of AI-Based Decision Support Systems in the Clinic,' *Journal of Medical Ethics*, 3 April 2020, medethics-2019-105860, https://doi.org/10.1136/medethics-2019-105860; Mark Coeckelbergh, 'Artificial Intelligence, Responsibility Attribution, and a Relational Justification of Explainability,' *Science and Engineering Ethics* 7, no. 0123456789 (2019), https://doi.org/10.1007/s11948-019-00146-8.

10 Herbert Lionel Adolphus Hart, *Punishment and Responsibility* (Oxford: Oxford University Press, 1968); Joel Feinberg, *Doing & Deserving; Essays in the Theory of Responsibility* (Princeton, NJ: Princeton University Press, 1970).

11 See Filippo Santoni de Sio and Ezio Di Nucci, 'Drones and Responsibility: Mapping the Field,' in *Drones and Responsibility: Legal, Philosophical and Socio-Technical Perspectives on Remotely Controlled Weapons*, eds. Ezio Di Nucci and Filippo Santoni de Sio (London and New York: Routledge, 2016) for an early attempt to clarify this.

12 H. L. A. Hart, *Punishment and Responsibility* (New York and Toronto: Oxford University Press, 1968); David Collingridge, *The Social Control of Technology* (London: Frances Printers, 1980); Mark Bovens, *The Quest for Responsibility: Accountability and Citizenship in Complex Organisations* (Cambridge: Cambridge University Press, 1998); Udo Pesch, 'Engineers and Active Responsibility,' *Science and Engineering Ethics* 21, no. 4 (9 August 2015): 925–39, https://doi.org/10.1007/s11948-014-9571-7; Ibo van de Poel and Martin Sand, 'Varieties of Responsibility: Two Problems of Responsible

Innovation,' *Synthese* 198, no. Suppl 190 (2021): 4769–87, https://doi.org/10.1007/s11229-018-01951-7.

13 Matthias, 'The Responsibility Gap,' 175–83; Sparrow, 'Killer Robots.'

14 Ugo Pagallo, *The Laws of Robots: Crimes, Contracts, and Torts* (Dordrecht: Springer, 2013); Ryan Calo, 'Robotics and the Lessons of Cyberlaw,' *California Law Review* 103, no. 3 (2015): 513–63, https://doi.org/10.2139/ssrn.2402972.

15 C. Heyns, *Report of the Special Rapporteur on Extra-Judicial, Summary or Arbitrary Executions* (Geneva: Human Rights, 2013); Chantal Meloni, 'State and Individual Responsibility for Targeted Killings by Drones,' in *Drones and Responsibility: Legal, Philosophical and Socio-Technical Perspectives on Remotely Controlled Weapons*, eds. Ezio Di Nucci and Filippo Santoni de Sio (London and New York: Routledge, 2016), 47–64.

16 Mittelstadt, Brent Daniel, Patrick Allo, Mariarosaria Taddeo, Sandra Wachter, and Luciano Floridi, 'The Ethics of Algorithms: Mapping the Debate,' *Big Data & Society* 3, no. 2 (December 2016): 1–21, https://doi.org/10.1177/2053951716679679; Derek Doran, Sarah Schulz, and Tarek R. Besold, 'What Does Explainable AI Really Mean? A New Conceptualization of Perspectives,' *CEUR Workshop Proceedings* 2071 (2 October 2017), https://arxiv.org/abs/1710.00794v1; Frank Pasquale, *The Black Box Society* (Cambridge, MA and London: Harvard University Press, 2015).

17 Matthias, 'The Responsibility Gap'; Sparrow, 'Killer Robots.'

18 Thomas W. Simpson and Vincent C. Müller, 'Just War and Robots' Killings,' *The Philosophical Quarterly* 66, no. 263 (1 April 2016): 302–22, https://doi.org/10.1093/pq/pqv075.

19 Jack Stilgoe, 'Machine Learning, Social Learning and the Governance of Self-Driving Cars,' *Social Studies of Science* 48, no. 1 (2017): 25–56, https://doi.org/10.1177/0306312717741687.

20 Hart, *Punishment and Responsibility*.

21 John Gardner, 'The Mark of Responsibility,' in *Offences and Defences*, ed. John Gardner (Oxford and New York: Oxford University Press, 2007), 177–200, https://doi.org/10.1093/acprof:oso/9780199239351.003.0009.

22 Bovens, *The Quest for Responsibility: Accountability and Citizenship in Complex Organisations*.

23 Ibo van de Poel, Lambèr M. M. Royakkers, and Sjoerd D. Zwart, *Moral Responsibility and the Problem of Many Hands* (London and New York: Routledge, 2015), https://doi.org/10.4324/9781315734217.

24 For a general discussion along these lines, see, e.g. D. Pereboom, *Living Without Free Will* (Cambridge: Cambridge University Press, 2001).

25 Stephen J. Morse, 'Moral and Legal Responsibility and the New Neuroscience,' in *Neuroethics in the 21st Century: Defining the Issues in Theory, Practice and Policy*, ed. Judy Illes (New York: Oxford University Press, 2006), 33.

26 P. F. Strawson, 'Freedom and Resentment,' *Proceedings of the British Academy* 48 (1962): 1–25.

27 Maureen Sie, *Justifying Blame: Why Free Will Matters and Why It Does Not* (Amsterdam: Rodopi, 2005).

28 Joel Feinberg, 'The Expressive Function of Punishment,' *The Monist* 49, no. 3 (1965): 397–423, https://doi.org/10.5840/monist196549326.

29 Matthias, 'The Responsibility Gap'; Sparrow, 'Killer Robots.'

30 Peter Asaro, 'A Body to Kick, but Still No Soul to Damn: Legal Perspectives on Robotics,' in *Robot Ethics*, eds. Patrick Lin, Keith Abney, and George A. Bekey (Cambridge, MA: The MIT Press, 2012), 169–86.

31 As a broader philosophical point, it has been argued that even cultural systems can be seen as types of 'automata', for their evolution cannot be controlled by individuals while rather shaping and constraining individual behaviour. See Carlo Sini, 'Machine, Culture, and Robot,' in *Italian Philosophy of Technology*, eds. Simona Chiodo and Viola Schiaffonati (Cham: Springer, 2021), 83–88.
32 Dennis F. Thompson, 'Moral Responsibility of Public Officials : The Problem of Many Hands,' *The American Political Science Review* 74, no. 4 (1980): 905–16; Bovens, *The Quest for Responsibility: Accountability and Citizenship in Complex Organisations*.
33 More on this in Section 7.4.1 on fatalism and deflationism.
34 John Danaher, 'Robots, Law and the Retribution Gap,' *Ethics and Information Technology* 18 (2016): 299–309, https://doi.org/10.1007/s10676-016-9403-3.
35 Susan Wolf, *Freedom within Reason* (New York: Oxford University Press, 1990); Michael McKenna, *Conversation & Responsibility* (Oxford: Oxford University Press, 2012).
36 Gardner, 'The Mark of Responsibility.'
37 Tony Honoré, *Responsibility and Fault* (Oxford: Hart Publishing, 1999).
38 Gardner, 'The Mark of Responsibility.'
39 Audley Genus and Andy Stirling, 'Collingridge and the Dilemma of Control: Towards Responsible and Accountable Innovation,' *Research Policy* 47 (2018): 61–69, https://doi.org/10.1016/j.respol.2017.09.012.
40 David Collingridge, *The Social Control of Technology* (London: Frances Printers, 1980) 62.
41 Charles E. Lindblom, *Inquiry and Change: The Troubled Attempt to Understand and Shape Society* (New Haven, CT and London: Yale University Press, 1990).
42 Jack Stilgoe, Richard Owen, and Phil Macnaghten, 'Developing a Framework for Responsible Innovation,' *Research Policy* 42, no. 9 (2013): 1568–80, https://doi.org/10.1016/j.respol.2013.05.008. See also Chapters 3 and 9 for a discussion of the Responsible Innovation approach.
43 Davide Castelvecchi, 'Can We Open the Black Box of AI?,' *Nature* 538, no. 7623 (October 2016): 20–23, https://doi.org/10.1038/538020a.
44 Frank Pasquale, *The Black Box Society* (Cambridge, MA and London: Harvard University Press, 2015), https://doi.org/10.4159/harvard.9780674736061.
45 Frank Pasquale, *The Black Box Society*.
46 Mark Bovens, 'Analysing and Assessing Accountability: A Conceptual Framework,' *European Law Journal* 13, no. 4 (2007): 447–68, '450'. Bovens just speaks of 'accountability,' but it is clear from the context of his work that he is talking of public accountability. I make the specification explicit to distinguish this from 'moral accountability' discussed above.
47 Bovens, *The Quest for Responsibility: Accountability and Citizenship in Complex Organisations*; Bovens, 'Analysing and Assessing Accountability: A Conceptual Framework.'
48 Guido Noto La Diega, 'Against the Dehumanisation of Decision-Making. Algorithmic Decisions at the Crossroads of Intellectual Property, Data Protection, and Freedom of Information,' *Journal of Intellectual Property, Information Technology and Electronic Commerce Law* 9, no. 1 (31 May 2018), https://doi.org/10.31228/osf.io/s2jnk; Mireille Hildebrandt, 'Privacy as Protection of the Incomputable Self: From Agnostic to Agonistic Machine Learning,' *Theoretical Inquiries in Law* 20 (23 January 2019), https://doi.org/10.1515/til-2019-0004; European Commission, 'High-Level Expert Group on Artificial Intelligence Ethics Guidelines for Trustworthy AI,' 2019, https://digital-strategy.ec.europa.eu/en/library/ethics-guidelines-trustworthy-ai.

49 Sandra Wachter, Brent Mittelstadt, and Luciano Floridi, 'Why a Right to Explanation of Automated Decision-Making Does Not Exist in the General Data Protection Regulation,' *International Data Privacy Law* 7, no. 2 (1 May 2017): 76–99, https://doi.org/10.1093/idpl/ipx005; Lilian Edwards and Michael Veale, 'Slave to the Algorithm? Why a Right to Explanation Is Probably Not the Remedy You Are Looking For,' *Duke Law and Technology Review* 16, no. 1 (2017): 1–65, https://doi.org/10.2139/ssrn.2972855.
50 Castelvecchi, 'Can We Open the Black Box of AI?'
51 Noto La Diega, 'Against the Dehumanisation of Decision-Making. Algorithmic Decisions at the Crossroads of Intellectual Property, Data Protection, and Freedom of Information.'
52 Stavros Zouridis, Marlies Van Eck, and Mark Bovens, 'Automated Discretion,' in *Discretion and the Quest for Controlled Freedom*, eds. Tony Evans and Peter Hupe (London and New York: Palgrave MacMillan, 2019).
53 Michael Lipsky, *Street-Level Bureaucracy: Dilemmas of the Individual in Public Services* (New York: Russell Sage Foundation, 1980); Mark Bovens and Stavros Zouridis, 'From Street-Level to System-Level Bureaucracies: How Information and Communication Technology Is Transforming Administrative Discretion and Constitutional Control,' *Public Administration Review* 62, no. 2 (January 2002): 174–84, https://doi.org/10.1111/0033-3352.00168.
54 Zouridis, Bovens, and Van Eck, 'Digital Discretion.'
55 Marlies Van Eck, 'Geautomatiseerde Ketenbesluiten & Rechtsbescherming: Een Onderzoek Naar de Praktijk van Geautomatiseerde Ketenbesluiten over Een Financieel Belang in Relatie Tot Rechtsbescherming,' (Tilburg Law School, 2018), https://research.tilburguniversity.edu/en/publications/automated-administrative-chain-decisions-amp-legal-protection-res.
56 Pasquale, *The Black Box Society*.
57 Virginia Eubanks, *Automating Inequality: How High-Tech Tools Profile, Police, and Punish the Poor* (New York: St. Martin's Press, 2018).
58 Thompson, 'Moral Responsibility of Public Officials: The Problem of Many Hands'; Bovens, *The Quest for Responsibility: Accountability and Citizenship in Complex Organisations*.
59 Bovens, *The Quest for Responsibility: Accountability and Citizenship in Complex Organisations*.
60 Tsjalling Swierstra and Jaap Jelsma, 'Responsibility without Moralism in Technoscientific Design Practice,' *Science, Technology, & Human Values* 31, no. 3 (19 May 2006): 309–32, https://doi.org/10.1177/0162243905285844.
61 Ibo van de Poel, Lambèr Royakkers, and Sjoerd Zwart, *Moral Responsibility and the Problem of Many Hands, Moral Responsibility and the Problem of Many Hands* (London and New York: Routledge, 2015), https://doi.org/10.4324/9781315734217.
62 Pesch, 'Engineers and Active Responsibility,' 9 August 2015.
63 Udo Pesch, 'Engineers and Active Responsibility,' *Science and Engineering Ethics* 21, no. 4 (9 August 2015): 930, https://doi.org/10.1007/s11948-014-9571-7.
64 A remarkable example of this attitude is that of some of the IT developers who invented popular social network platforms such as Facebook and Twitter, interviewed in the Netflix documentary *The Social Dilemma* (2020).
65 Van de Poel and Sand, 'Varieties of Responsibility: Two Problems of Responsible Innovation.'
66 Matthias, 'The Responsibility Gap.'
67 Matthias.
68 Sparrow, 'Killer Robots.'

69 Filippo Santoni de Sio, 'Killing by Autonomous Vehicles and the Legal Doctrine of Necessity,' *Ethical Theory and Moral Practice* (2017): 411–29, https://doi.org/10.1007/s10677-017-9780-7.
70 Madeleine Hayenhjelm and Jonathan Wolff, 'The Moral Problem of Risk Impositions: A Survey of the Literature,' *European Journal of Philosophy* 20 (June 2012): E26–51, https://doi.org/10.1111/j.1468-0378.2011.00482.x.
71 Simpson and Müller, 'Just War and Robots' Killings.'
72 In the exact words of Simpson and Müller: 'If it is possible to build robots that comply with a strict tolerance level, then that is obligatory. But if it is not possible, the best we can do is good enough,' 316–17.
73 Ryan Calo, 'Robotics and the Lessons of Cyberlaw'; Pagallo, *The Laws of Robots: Crimes, Contracts, and Torts.*
74 In fact, such culpability gaps in the form of the 'problem of many hands' may also occur in contexts where AI is not involved. See van de Poel, Royakkers, and Zwart, *Moral Responsibility and the Problem of Many Hands.*
75 This paragraph is based on the story of the Bolam v Bolitho legal standard of negligence in the British law, and its reconstruction by Rachael Mulheron, 'Trumping Bolam: A Critical Legal Analysis of Bolitho's 'Gloss,'' *The Cambridge Law Journal* 69, no. 3 (12 November 2010): 609–38, https://doi.org/10.1017/S0008197310000826.
76 S. O. Funtowicz and J. R. Ravetz, 'Post-Normal Science: A New Science for New Times, October 1990,' *Scientific European* (1990): 20–22.
77 Jeroen van den Hoven, Gert Jan Lokhorst, and Ibo van de Poel, 'Engineering and the Problem of Moral Overload,' *Science and Engineering Ethics* 18, no. 1 (2012): 143–55, https://doi.org/10.1007/s11948-011-9277-z.
78 Evgeny Morozov, *To Save Everything, Click Here: The Folly of Technological Solutionism* (New York: Public Affairs, 2013); Stilgoe, 'Machine Learning, Social Learning and the Governance of Self-Driving Cars.'
79 Derek Doran, Sarah Schulz, and Tarek R. Besold, 'What Does Explainable AI Really Mean? A New Conceptualization of Perspectives,' *CEUR Workshop Proceedings* 2071 (2 October 2017).
80 Doran, Schulz, and Besold.
81 Brent Daniel Mittelstadt et al., 'The Ethics of Algorithms: Mapping the Debate.'
82 Mittelstadt et al., 12.
83 Judith Simon, 'Distributed Epistemic Responsibility in a Hyperconnected Era,' in *The Onlife Manifesto*, ed. Luciano Floridi (Cham: Springer International Publishing, 2015), 145–59, https://doi.org/10.1007/978-3-319-04093-6_17.
84 Van de Poel, Royakkers, and Zwart, *Moral Responsibility and the Problem of Many Hands.*
85 Noto La Diega, 'Against the Dehumanisation of Decision-Making. Algorithmic Decisions at the Crossroads of Intellectual Property, Data Protection, and Freedom of Information.'
86 See the discussion of the 'tracing condition' for Meaningful Human Control in Chapter 8 below.
87 Calo, 'Robotics and the Lessons of Cyberlaw'; Ugo Pagallo, *The Laws of Robots: Crimes, Contracts, and Torts.*
88 Maurice Schellekens, 'No-Fault Compensation Schemes for Self-Driving Vehicles,' *Law, Innovation and Technology* 10, no. 2 (3 July 2018): 314–33, https://doi.org/10.1080/17579961.2018.1527477.
89 Mady Delvaux, 'Report with Recommendations to the Commission on Civil Law Rules on Robotics (A8-0005/2017),' 2017, https://www.europarl.europa

.eu/doceo/document/A-8-2017-0005_EN.html#title1; Bert-Jaap Koops, Mireille Hildebrandt, and David-Olivier Jaquet-Chiffelle, 'Bridging the Accountability Gap: Rights for New Entities in the Information Society,' *Minnesota Journal of Law, Science and Technology* 11, no. 2 (2010): 497–561; Joanna J. Bryson, Mihailis E. Diamantis, and Thomas D. Grant, 'Of, for, and by the People: The Legal Lacuna of Synthetic Persons,' *Artificial Intelligence and Law* 25, no. 3 (8 September 2017): 273–91, https://doi.org/10.1007/s10506-017-9214-9.

90 Ibo van de Poel et al., 'The Problem of Many Hands: Climate Change as an Example,' *Science and Engineering Ethics* 18, no. 1 (2012): 49–67, https://doi.org/10.1007/s11948-011-9276-0.

91 A prominent example is the recent story of the European data protection legislation GDPR that some critics summarised as the 'world's toughest privacy law' proven 'toothless'. See e.g. Nicholas Vinocur, "We Have a Huge Problem': European Tech Regulator Despairs over Lack of Enforcement,' *Politico* (27 Dec 2019), https://www.politico.eu/article/we-have-a-huge-problem-european-regulator-despairs-over-lack-of-enforcement/

References

Asaro, Peter. 'A Body to Kick, But Still No Soul to Damn: Legal Perspectives on Robotics.' In *Robot Ethics*, edited by Patrick Lin, Keith Abney, and George A Bekey, 169–86. Cambridge, MA: The MIT Press, 2012.

Bovens, Mark. 'Analysing and Assessing Accountability: A Conceptual Framework.' *European Law Journal* 13, no. 4 (2007): 447–68.

———. *The Quest for Responsibility: Accountability and Citizenship in Complex Organisations*. Cambridge: Cambridge University Press, 1998.

Bovens, Mark, and Stavros Zouridis. 'From Street-Level to System-Level Bureaucracies: How Information and Communication Technology Is Transforming Administrative Discretion and Constitutional Control.' *Public Administration Review* 62, no. 2 (January 2002): 174–84. https://doi.org/10.1111/0033-3352.00168.

Braun, Matthias, Patrik Hummel, Susanne Beck, and Peter Dabrock. 'Primer on an Ethics of AI-Based Decision Support Systems in the Clinic.' *Journal of Medical Ethics* (3 April 2020). https://doi.org/10.1136/medethics-2019-105860.

Bryson, Joanna J., Mihailis E. Diamantis, and Thomas D. Grant. 'Of, for, and by the People: The Legal Lacuna of Synthetic Persons.' *Artificial Intelligence and Law* 25, no. 3 (8 September 2017): 273–91. https://doi.org/10.1007/s10506-017-9214-9.

Calo, Ryan. 'Robotics and the Lessons of Cyberlaw.' *California Law Review* 103, no. 3 (2015): 513–63. https://doi.org/10.2139/ssrn.2402972.

Castelvecchi, Davide. 'Can We Open the Black Box of AI?' *Nature* 538, no. 7623 (October 2016): 20–23. https://doi.org/10.1038/538020a.

Coeckelbergh, Mark. 'Artificial Intelligence, Responsibility Attribution, and a Relational Justification of Explainability.' *Science and Engineering Ethics* 7, no. 0123456789 (2019). https://doi.org/10.1007/s11948-019-00146-8.

Collingridge, David. *The Social Control of Technology*. London: Frances Printers, 1980.

Danaher, John. 'Robots, Law and the Retribution Gap.' *Ethics and Information Technology* 18 (2016): 299–309. https://doi.org/10.1007/s10676-016-9403-3.

Delvaux, Mady. 'Report with Recommendations to the Commission on Civil Law Rules on Robotics (A8-0005/2017).' 2017. https://www.europarl.europa.eu/doceo/document/A-8-2017-0005_EN.html#title1.

Dennett, Daniel C. 'When Hal Kills, Who's to Blame? Computer Ethics.' In *Hal's Legacy: 2001's Computer as Dream and Reality*, edited by David D. Stork, 351–65. Cambridge, MA: The MIT Press, 1997.

Doran, Derek, Sarah Schulz, and Tarek R. Besold. 'What Does Explainable AI Really Mean? A New Conceptualization of Perspectives.' *CEUR Workshop Proceedings* 2071 (2 October 2017). https://arxiv.org/abs/1710.00794v1.

Edwards, Lilian, and Michael Veale. 'Slave to the Algorithm? Why a Right to Explanationn Is Probably Not the Remedy You Are Looking For.' *Duke Law and Technology Review* 16, no. 1 (2017): 1–65. https://doi.org/10.2139/ssrn.2972855.

Eubanks, Virginia. *Automating Inequality: How High-Tech Tools Profile, Police, and Punish the Poor*. New York: St. Martin's Press, 2018.

European Commission. 'High-Level Expert Group on Artificial Intelligence Ethics Guidelines for Trustworthy AI.' 2019. https://digital-strategy.ec.europa.eu/en/library/ethics-guidelines-trustworthy-ai.

Feinberg, Joel. *Doing & Deserving; Essays in the Theory of Responsibility*. Princeton: Princeton University Press, 1970.

———. 'The Expressive Function of Punishment,' *The Monist* 49, no. 3 (1965): 397–423. https://doi.org/10.5840/monist196549326.

Funtowicz, S. O., and J. R. Ravetz. 'Post-Normal Science: A New Science for New Times.' *Scientific European* (October 1990): 20–22.

Gardner, John. 'The Mark of Responsibility.' In J. Gardner, *Offences and Defences*, 177–200. New York: Oxford University Press, 2007. https://doi.org/10.1093/acprof:oso/9780199239351.003.0009.

Genus, Audley, and Andy Stirling. 'Collingridge and the Dilemma of Control: Towards Responsible and Accountable Innovation.' *Research Policy* 40, 47: (2018): 61–69. https://doi.org/10.1016/j.respol.2017.09.012.

Hart, Herbert Lionel Adolphus. *Punishment and Responsibility*. New York and Toronto: Oxford University Press, 1968. http://www.oxfordscholarship.com/view/10.1093/acprof:oso/9780199534777.001.0001/acprof-9780199534777.

Hayenhjelm, Madeleine, and Jonathan Wolff. 'The Moral Problem of Risk Impositions: A Survey of the Literature.' *European Journal of Philosophy* 20 (June 2012): E26–E51. https://doi.org/10.1111/j.1468-0378.2011.00482.x.

Heyns, C. *Report of the Special Rapporteur on Extra-Judicial, Summary or Arbitrary Executions*. Geneva: Human Rights Council, 2013.

Hildebrandt, Mireille. 'Privacy as Protection of the Incomputable Self: From Agnostic to Agonistic Machine Learning.' *Theoretical Inquiries in Law* 20 (23 January 2019). https://doi.org/10.1515/til-2019-0004.

Honoré, Tony. *Responsibility and Fault*. Oxford: Hart Publishing, 1999.

Human Right Watch. 'Mind the Gap: The Lack of Accountability for Killer Robots.' 2015. https://www.hrw.org/report/2015/04/09/mind-gap/lack-accountability-killer-robots

Koops, Bert-Jaap, Mireille Hildebrandt, and David-Olivier Jaquet-Chiffelle. 'Bridging the Accountability Gap: Rights for New Entities in the Information Society.' *Minnesota Journal of Law, Science and Technology* 11 (2010): 497–561.

Lindblom, Charles E. *Inquiry and Change : The Troubled Attempt to Understand and Shape Society*. New Haven, CT and London: Yale University Press, 1990.

Lipsky, Michael. *Street-Level Bureaucracy : Dilemmas of the Individual in Public Services*. New York: Russell Sage Foundation, 1980.

Matthias, Andreas. 'The Responsibility Gap: Ascribing Responsibility for the Actions of Learning Automata.' *Ethics and Information Technology* 6, no. 3 (1 September 2004): 175–83. https://doi.org/10.1007/s10676-004-3422-1.

McKenna, Michael. *Conversation & Responsibility*. New York: Oxford University Press, 2012.

Meloni, Chantal. 'State and Individual Responsibility for Targeted Killings by Drones.' In *Drones and Responsibility: Legal, Philosophical and Socio-Technical Perspectives on Remotely Controlled Weapons*, edited by Ezio Di Nucci and Filippo Santoni de Sio, 47–64. London and New York: Routledge, 2016.

Mittelstadt, Brent Daniel, Patrick Allo, Mariarosaria Taddeo, Sandra Wachter, and Luciano Floridi. 'The Ethics of Algorithms: Mapping the Debate.' *Big Data & Society* 3, no. 2 (December 2016): 1–21. https://doi.org/10.1177/2053951716679679.

Morozov, Evgeny. *To Save Everything, Click Here: The Folly of Technological Solutionism*. New York: Public Affairs, 2013.

Morse, Stephen J. 'Moral and Legal Responsibility and the New Neuroscience.' In *Neuroethics in the 21st Century: Defining the Issues in Theory, Practice and Policy 33*, edited by Judy Illes. Oxford: Oxford University Press, 2006.

Mulheron, Rachael. 'Trumping Bolam : A Critical Legal Analysis of Bolitho's "Gloss".' *The Cambridge Law Journal* 69, no. 3 (12 November 2010): 609–38. https://doi.org/10.1017/S0008197310000826.

Noto La Diega, Guido. 'Against the Dehumanisation of Decision-Making. Algorithmic Decisions at the Crossroads of Intellectual Property, Data Protection, and Freedom of Information.' *Journal of Intellectual Property, Information Technology and Electronic Commerce Law* 9, no. 1 (31 May 2018). https://doi.org/10.31228/osf.io/s2jnk.

Nyholm, Sven. 'Attributing Agency to Automated Systems: Reflections on Human–Robot Collaborations and Responsibility-Loci.' *Science and Engineering Ethics* 24, no. 4 (18 August 2018): 1201–19. https://doi.org/10.1007/s11948-017-9943-x.

Pagallo, Ugo. *The Laws of Robots: Crimes, Contracts, and Torts*. Dordrecht: Springer, 2013. http://www.springer.com/gp/book/9789400765634.

Pasquale, Frank. *The Black Box Society*. Cambridge, MA and London: Harvard University Press, 2015. https://doi.org/10.4159/harvard.9780674736061.

Pereboom, Derk. *Living Without Free Will*. Cambridge: Cambridge University Press, 2001.

Pesch, Udo. 'Engineers and Active Responsibility.' *Science and Engineering Ethics* 21, no. 4 (9 August 2015): 925–39. https://doi.org/10.1007/s11948-014-9571-7.

Santoni de Sio, Filippo, and Ezio Di Nucci. 'Drones and Responsibility: Mapping the Field.' In *Drones and Responsibility: Legal, Philosophical, and Socio-Technical Perspectives on the Use of Remotely Controlled Weapons*, edited by Ezio Di Nucci and Filippo Santoni de Sio, 1–14. London and New York: Routledge, 2016.

Santoni de Sio, Filippo. 'Killing by Autonomous Vehicles and the Legal Doctrine of Necessity.' *Ethical Theory and Moral Practice* (2017): 411–29. https://doi.org/10.1007/s10677-017-9780-7.

Santoni de Sio, Filippo, and Giulio Mecacci. 'Four Responsibility Gaps with Artificial Intelligence: Why They Matter and How to Address Them.' *Philosophy & Technology* 34, no. 4 (14 December 2021): 1057–84. https://doi.org/10.1007/s13347-021-00450-x.

Schellekens, Maurice. 'No-Fault Compensation Schemes for Self-Driving Vehicles.' *Law, Innovation and Technology* 10, no. 2 (3 July 2018): 314–33. https://doi.org/10.1080/17579961.2018.1527477.

Sie, Maureen. *Justifying Blame: Why Free Will Matters and Why It Does Not.* Amsterdam: Rodopi, 2005.

Simon, Judith. 'Distributed Epistemic Responsibility in a Hyperconnected Era.' In *The Onlife Manifesto*, edited by Luciano Floridi, 145–59. Cham: Springer, 2015. https://doi.org/10.1007/978-3-319-04093-6_17.

Simpson, Thomas W., and Vincent C Müller. 'Just War and Robots' Killings.' *The Philosophical Quarterly* 66, no. 263 (1 April 2016): 302–22. https://doi.org/10.1093/pq/pqv075.

Sini, Carlo. 'Machine, Culture, and Robot.' In *Italian Philosophy of Technology*, edited by Simona Chiodo and Viola Schiaffonati, 83–88. Cham: Springer, 2021.

Smiley, Lauren. 'The Legal Saga of Uber's Fatal Self-Driving Car Crash Is Over.' *Wired*, accessed 16 October 2023. https://www.wired.com/story/ubers-fatal-self-driving-car-crash-saga-over-operator-avoids-prison/.

Sparrow, Robert. 'Killer Robots.' *Journal of Applied Philosophy* 24, no. 1 (2007): 62–77. https://doi.org/10.1111/j.1468-5930.2007.00346.x.

Stilgoe, Jack. 'Machine Learning, Social Learning and the Governance of Self-Driving Cars.' *Social Studies of Science* 48, no. 1: 25–56. https://doi.org/10.1177/0306312717741687.

———. 'Who Killed Elaine Herzberg?' In Jack Stilgoe, *Who's Driving Innovation? New Technologies and the Collaborative State*, 1–6. Cham: Palgrave Macmillan, 2020. https://doi.org/10.1007/978-3-030-32320-2_1.

Stilgoe, Jack, Richard Owen, and Phil Macnaghten. 'Developing a Framework for Responsible Innovation.' *Research Policy* 42, no. 9 (2013): 1568–80. https://doi.org/10.1016/j.respol.2013.05.008.

Strawson, P. F. 'Freedom and Resentment.' *Proceedings of the British Academy* 48 (1962): 1–25.

Swierstra, Tsjalling, and Jaap Jelsma. 'Responsibility without Moralism in Technoscientific Design Practice.' *Science, Technology, & Human Values* 31, no. 3 (19 May 2006): 309–32. https://doi.org/10.1177/0162243905285844.

Thompson, Dennis F. 'Moral Responsibility of Public Officials : The Problem of Many Hands.' *The American Political Science Review* 74, no. 4 (1980): 905–16.

Van de Poel, Ibo, Jessica Nihlén Fahlquist, Neelke Doorn, Sjoerd Zwart, and Lambèr Royakkers. 'The Problem of Many Hands: Climate Change as an Example.' *Science and Engineering Ethics* 18, no. 1 (2012): 49–67. https://doi.org/10.1007/s11948-011-9276-0.

Van de Poel, Ibo, Lambèr Royakkers, and Sjoerd Zwart. *Moral Responsibility and the Problem of Many Hands*. London and New York: Routledge, 2015. https://doi.org/10.4324/9781315734217.

Van de Poel, Ibo, and Martin Sand. 'Varieties of Responsibility: Two Problems of Responsible Innovation.' *Synthese* 198, no. Suppl 190 (2021): 4769–87. https://doi.org/10.1007/s11229-018-01951-7.

Van den Hoven, Jeroen, Gert Jan Lokhorst, and Ibo van de Poel. 'Engineering and the Problem of Moral Overload.' *Science and Engineering Ethics* 18, no. 1 (2012): 143–55. https://doi.org/10.1007/s11948-011-9277-z.

Van Eck, Marlies. 'Geautomatiseerde Ketenbesluiten & Rechtsbescherming: Een Onderzoek Naar de Praktijk van Geautomatiseerde Ketenbesluiten over Een Financieel Belang in Relatie Tot Rechtsbescherming.' Tilburg Law School, 2018. https://research.tilburguniversity.edu/en/publications/automated-administrative-chain-decisions-amp-legal-protection-res.

Vinocur, Nicholas. '"We Have a Huge Problem": European Tech Regulator Despairs over Lack of Enforcement.' *Politico* (27 Dec 2019). https://www.politico.eu/article/we-have-a-huge-problem-european-regulator-despairs-over-lack-of-enforcement/.

Wachter, Sandra, Brent Mittelstadt, and Luciano Floridi. 'Why a Right to Explanation of Automated Decision-Making Does Not Exist in the General Data Protection Regulation.' *International Data Privacy Law* 7, no. 2 (1 May 2017): 76–99. https://doi.org/10.1093/idpl/ipx005.

Wolf, Susan. *Freedom within Reason*. New York: Oxford University Press, 1990.

Zouridis, Stavros, Marlies Van Eck, and Mark Bovens. 'Digital Discretion.' In *Discretion and the Quest for Controlled Freedom*, edited by Tony Evans and Peter Hupe. London and New York: Palgrave Macmillan, 2019.

8 Meaningful Human Control over AI

8.1 Stopkillerrobots.org

If, on a spring morning in 2013, you walked by the Houses of Parliament in London, you could have spotted a group of people standing next to a giant toy robot and holding a big banner with the slogan: *Stop Killer Robots*. They were people from NGOs as well as scientists, roboticists, and lawyers who had launched a political campaign for an international ban on so-called 'autonomous weapon systems.' AI-driven autonomous weapon systems are systems that 'once launched will select and engage targets without further human intervention.'[1] Far from being science-fictional dreams or nightmares, the development of such systems by various countries has reportedly been going on since as early as 2014.[2] The Super aEgis II, a Korean-produced gun turret able to identify, track, and shoot targets without human intervention within a four-kilometre range, is available on the market and already being used – though not in the fully automated mode – in numerous locations in the Middle East.[3] Drones were decisive in the success of Azerbaijan's military attacks in the contested region of Nagorno Karabakh in 2020.[4] A report from the UN Security Council suggests that autonomous weapon systems may have been used in the recent civil war in Libya,[5] and media reports claim armed drones operating without human oversight may have been deployed by Ukraine in their defence operations against the Russian invasion.[6] Some defence experts believe that Ukraine has become a 'living lab for AI warfare',[7] and the Israeli Defence Force has allegedly used an AI-based system in the bombing of Gaza following Hamas' attacks of October 7 2023 to 'produce targets at fast pace' 'through the rapid and automatic extraction of intelligence.'[8]

Autonomous weapon systems have raised many ethical and political concerns: if they get out of hand, they could increase war crimes and wrongs, lower the threshold for waging war, violate human dignity, and create responsibility and accountability gaps for military operations.[9] The 'Stop Killer Robots' campaign raised a big public and political debate and contributed to prompt concrete political initiatives. The most remarkable

DOI: 10.4324/9781003303244-11

one was the organisation by the United Nations of a series of expert meetings in Geneva where representatives of States and civil societies – NGOs and researchers – discussed the proposal to preemptively ban the use of the so-called lethal autonomous weapon systems. Since 2018, United Nations Secretary-General António Guterres had endorsed the claim that lethal autonomous weapon systems are morally and politically unacceptable and called for their prohibition under international law. However, parties in the debate still tended to disagree on the definition of what should count as a 'fully autonomous system,' whether the term made any sense at all and whether a ban on their development and use would be the right way to address the above-mentioned risks.[10]

An important turn in this political debate was triggered in 2013 by the NGO Article 36, with the introduction in a series of reports of a term that would shape the debate for the years to come: 'Meaningful Human Control'. In a nutshell, the idea is that – no matter how intelligent and equipped with autonomous capabilities the weapon systems are – humans, not computers and their algorithms, should ultimately remain in control of, and thus morally responsible for, relevant decisions about (lethal) military operations.[11] This idea has attracted a wide consensus among scholars and policymakers.[12] The requirement of Meaningful Human Control effectively captured the fact that a mere human presence in the system, or the passive capacity to assess the behaviour of the system, would not be enough to give human actors an effective capacity and opportunity to have the system behave in desirable ways[13] and to remain responsible for each individual attack.[14]

The ethical and political debates on the use of AI systems in the military are still ongoing.[15] Multiple definitions of Meaningful Human Control over autonomous weapon systems have been proposed.[16]

8.2 The idea of Meaningful Human Control

The notion of Meaningful Human Control is philosophically promising as it is meant to capture three ideas. Firstly, *being in the loop* – as computer scientists sometimes say – does not equate *to being in control*. Simple human presence within a network of agents is not a sufficient condition for being in control of an activity mediated by AI because one can be present and able to influence some parts of the system while (a) not being able to influence other parts that could be even more relevant from a moral point of view than the parts one can influence and (b) not having enough information or options to influence the process, for instance, if the human task consists in 'merely pushing a button in a reflex when a light goes on.'[17] Secondly, *controlling* does not mean *being in control*. Controlling in the sense of being in the position of making a substantive causal contribution to an AI-mediated activity through intentional actions (for example, holding a remote control or sitting behind the steering wheel of an automated vehicle) might not be a sufficient condition for meaningful control either. This is the case, for instance, if the system does not allow an agent to intervene at the relevant time, or the agent does not have the psychological capacity to respond appropriately under the circumstances. Thirdly and relatedly, whereas some forms of legal responsibility – i.e. tort liability and strict liability – require only that the agent has relatively simple forms of causal control over events, other forms of legal responsibility – typically criminal responsibility – usually require stricter control conditions of knowledge, intention, capacity, and opportunity. Therefore, no matter how strong the political will is to keep some humans responsible or accountable for the behaviour of autonomous weapon systems, attributions of legal responsibility that are not grounded in the relevant control conditions may turn out to be not only morally unfair but also difficult to enforce in tribunals.[18] This is one of the dimensions of the problem of 'responsibility gap' discussed in the previous chapter (the 'culpability gap' with AI).

In fact, experts from different disciplines have recently converged towards the idea that granting Meaningful Human Control over AI would substantially contribute to addressing responsibility gaps beyond the defence context.[19] Clearly, the military is not the only place where AI may have a strong impact on human life (and death). Meaningful Human Control has been therefore recently investigated in the field of automated driving systems[20] and medical automation.[21] AI may impact human life also in a less direct and physical way, such as determining people's right to access (often vital) financial support, as in the infamous 'childcare benefit scandal' in the Netherlands, where thousands of families were mistakenly accused of fraud and forced to pay back large sums of money to the State, based on what turned out to be a flawed algorithmic system.[22] Relatedly, the importance as well as the difficulty of defining a kind of 'meaningful human involvement' as required to avoid responsibility gaps in automated decision-making is highlighted in art. 22(1) of the EU's GDPR as interpreted by the Article 29 Working Party/European Data Protection Board.[23]

8.3 The theory of Meaningful Human Control as reason-responsiveness

In this and the next chapter, I propose to look at the broader ethical and philosophical question: what should, more generally, be the role of human control over AI systems, such that unreasonable risks are avoided, human responsibility does not evaporate, and there is a place to turn to in case of untoward outcomes. My general ethical claim is that the presence of autonomous capabilities in AI systems can and should be combined with human control and responsibility. In the terms of the previous chapter, we should design systems to avoid responsibility gaps. Rather than focusing on one specific domain, or some specific legal or technical challenge to retain Meaningful Human Control, in this chapter I will present a general philosophical theory of Meaningful Human Control that identifies the generic conceptual and normative principles that should be considered to define, realise, and maintain Meaningful Human Control in specific contexts.[24]

This theory is grounded in general to the philosophical theories of moral responsibility and in particular to the notion of *guidance control* that has been developed by Fischer and Ravizza.[25] I adapt and expand this concept to cover actions involving the use of AI systems. The result is a theory of Meaningful Human Control, which, as I will show, can be fruitfully applied in the context of autonomous weapon systems as well as AI systems in general. To distinguish it from other approaches to Meaningful Human Control present in the literature, and in line with Fischer's and Ravizza's theory of responsibility from which it is inspired, I will refer to our theory as 'Meaningful Human Control as reason-responsiveness.'

The theory is also grounded in the ideal of Value-Sensitive Design,[26] according to which, in order to have a real impact on society, ethical constraints and aims should shape technology in the *design* phase, where they still can make a difference, instead of fuelling political and academic discussions when technology is already in place; therefore, they should eventually be cast in terms that can actually be designed for, i.e. in the form of requirements for systems, engineering, and software design.[27]

In this chapter, I will present the general lines of the theory of Meaningful Human Control as reason-responsiveness. The next chapter will delve into its social preconditions, in particular, the pre-requirement of 'epistemic justice' necessary to make Meaningful Human Control potentially available to all the people, and not only to those with a stronger social position. The remainder of this chapter proceeds as follows. I first briefly introduce the philosophical debate on control and moral responsibility and the distinction between incompatibilist and compatibilist theories of moral responsibility. I explain why I consider the compatibilist approaches to moral responsibility most suitable to ground a theory of Meaningful Human Control over AI systems (Section 8.4); I introduce Fischer & Ravizza's account of guidance control (Section 8.5), and I expand, integrate, and translate it into a theory of Meaningful Human Control over actions mediated by AI systems and a set of design requirements to achieve this kind of control (Section 8.6). I finally show the fruitfulness of our approach to Meaningful Human Control for the military domain and autonomous weapon systems (Section 8.7) and for designing controllable AI systems and preventing responsibility gaps more generally (Section 8.8).

8.4 Free will and moral responsibility: the compatibilist way

Let us first briefly recap a dispute in Western philosophy that has been going on for many centuries: the debate on the possibility of human moral responsibility. This debate focuses on the question of whether and under which conditions people are *in control of* and, therefore, *responsible for* their everyday actions. The so-called 'incompatibilists' believe that people can be in control of and morally responsible for their actions if and only if they possess a special power to make decisions and carry out actions that escape the causal influence of – depending on the flavour of the time – divine omnipotence, determinist laws of physics, or genetic, (neuro)biological, socio-psychological, and circumstantial factors. They are called incompatibilists because they deny the compatibility of causal explanations of human actions and human moral responsibility, stating that causality and moral responsibility cannot be reconciled. If human actions are causally determined by God, physics, genetics social factors, etc., then they cannot be free and responsible. Incompatibilists can be divided into two main groups, depending on which of the two incompatible notions they hold on to. Some are 'libertarians.'[28] In line with a philosophical tradition that goes back at least to Immanuel Kant, they believe that human persons possess a special kind of autonomy, a 'contra-causal' power which gives them a special metaphysical status and makes them morally responsible for their actions in a sense in which no other creature is or can be. Other incompatibilists are 'free-will sceptics,' who believe that humans are not free or autonomous in any special sense; that is, they do not possess any special power to escape the causal influences on their behaviour and, therefore, from a philosophical point of view, they are ultimately never morally responsible for their actions; human actions are, as it were, in no relevant way different than natural events. Free will and moral responsibility are an illusion.[29]

In contrast with both forms of incompatibilism, the so-called 'compatibilists' believe that humans may be morally responsible for (some of) their actions even if they do not possess any special metaphysical power to escape the causal influences on their behaviour. Traditional compatibilism of modern philosophers like Hobbes and Hume was grounded in a mechanistic and associationist view of the human mind. According to traditional compatibilism, for agents to be morally responsible, their actions need to be free only in the sense of being the causal product of internal motivational factors – desires, intentions, traits of characters, and values – as opposed to being the product of 'external' forces, that is the product of physical or psychological coercion. Freedom here just means the absence of coercion. Present-day compatibilists,[30] while also rejecting the idea of strong metaphysical free will as a necessary condition for moral responsibility, rely on a more complex view of the human mind and action. They reject the idea of mental causation as being sufficient to ground moral responsibility – many mentally caused actions are not responsible, for instance, those carried out by mentally disordered persons. Therefore, they add a stricter condition for free will and moral responsibility: the capacity for rational control over one's action.

Present-day compatibilism offers a very suitable ground to elaborate an account of Meaningful Human Control over AI systems. Unlike incompatibilist libertarians, compatibilists do not ground human moral responsibility in any special, supernatural, exclusively human contra-causal power, and therefore they are not committed to the claim that *any* delegation of decision-making to non-human agents amounts per se to a disappearance of human moral responsibility over decisions and actions.[31] Unlike free-will sceptics, however, compatibilists do recognise that there is a difference between human actions and other natural events, and they thus claim that human agents can be legitimately seen as morally responsible for at least some of their actions. Finally, unlike traditional compatibilists, present-day compatibilists try to define stricter requirements for the kind of control over actions required for moral responsibility, which do not simply coincide with the causal power to bring about effects in the world through one's desires and intentional actions.

In the next section, I focus on a very influential compatibilist theory of human control over everyday actions: Fischer and Ravizza's theory of 'guidance control.'[32] I do so as I think that Fischer and Ravizza's theory of control – whatever its merits as a theory of moral responsibility – is a very promising starting point for a theory of Meaningful Human Control over AI systems. In the subsequent sections, I apply a version of Fischer and Ravizza's theory to institutions and decision-making processes powered by AI systems.

8.5 Conditions for 'guidance control'

According to Fischer and Ravizza, to be morally responsible for an action, a person should possess what they call 'guidance control' over that action. Guidance control is realised when two conditions are met: the decisional mechanism leading up to the action should be (1) 'moderately reason-responsive,' and (2) the decisional mechanism should be 'the agent's own.'[33]

The first condition of the reason-responsiveness of the decisional mechanism requires that the person must act according to a decisional mechanism that, in the presence of strong reasons to act or to not act, can recognise these reasons and bring that person to (not) perform that action in a sufficiently broad range of circumstances. This requirement separates morally responsible actors from actors acting under excusing factors such as non-culpably being under the influence of potent drugs, direct manipulation of the brain, behaviour attributable to a brain lesion or a neurological disorder, phobias, addictions, and coercive threats.[34] In all these cases, the person's decisional mechanism is bypassed or not responsive enough to moral reasons: it would and does lead to actions even in the presence of strong contrary moral reasons recognised by the agents, or it fails to lead to certain actions even in the presence of strong moral reasons to act. The latter could be the case of an agoraphobic who does not venture onto the street to help someone who could be rescued without serious risk. Although the person acknowledges the validity of the moral reasons in favour of helping the person in the street, their phobia makes it extremely difficult for them to do what they recognise as the right thing to do.

According to Fischer and Ravizza's second condition for guidance control, to be morally responsible for their actions, the decisional mechanism should also be an integral part of who the agent is. This means that the agent must have 'taken responsibility' for the mechanism by which they decide. 'Taking responsibility' for one's decisional mechanism requires that (a) the agent sees that their decisions have certain effects in the world; (b) the agent sees that others may have moral reactions towards them because of how they affect the world; and (c) the views specified in the first two conditions are based on the agent's evidence in an appropriate way.[35] If you consciously decide to settle an important matter by the toss of a coin, you are aware that you have decided to rely on what you perfectly know to be a random decision-making process, and you must accept that others will hold you accountable for the outcome of the choice: you cannot blame the system of your own choice.

This second condition for guidance control, which may be called the 'ownership' condition, separates morally responsible actions from behaviour resulting from a reason-responsive mechanism that the agent has reasons *not* to consider as their own, like in cases of psychological manipulation, subliminal persuasion, strong nudging, strong entrapment, brainwashing, and indoctrination.

8.6 Tracking and tracing conditions

The theory of guidance control presents the conditions for human agents to be morally responsible for their everyday actions based on the features of the decisional mechanism leading to those actions, as well as the relation between the agent and the decision-making mechanism.

Fischer and Ravizza primarily consider intra-personal decision mechanisms, that is, the mechanisms of an individual human mind or brain; however, an influential body of work in present-day cognitive science and philosophy of the human mind claims that human decisions and actions and even human personality traits are not represented in any brain mechanism; these accounts describe the human mind as 'extended, embedded and embodied.'[36] Accordingly, the idea of human control over actions should be applied on a larger scale to include artefacts and engineering systems;[37] in fact, if we consider AI systems as part of the decision-making mechanisms through which human agents carry out actions in the world, then conditions for guidance control on everyday actions may provide the basis for an account of Meaningful Human Control over actions mediated by AI systems. In what follows, I offer an outline of such an account of Meaningful Human Control by elaborating, in turn, on each of the two conditions for guidance control presented by Fischer and Ravizza.

8.6.1 Tracking[38]

Applied to complex systems operated by humans, Fischer and Ravizza's first condition for guidance control – the *reason-responsiveness* condition – requires that the operator's decisional mechanism is sensitive and responsive to a sufficient variety of moral inputs, i.e. that it can adapt the behaviour of the system to the relevant moral features of the circumstances. These comprise both the mental states of human operators and features of the external world. What is required is that the behaviour of the combined system (the human operator(s) and the complex system, including interfaces that support decision-making) co-varies with the moral reasons of the human operators for (not) carrying out a certain action.

There are many cases where intelligent systems with autonomous capabilites have misrepresented the relevant states of affairs and, as a result, have not been able to behave in accordance with the relevant human reasons. Many aeroplane crashes are the result of erroneous sensor data and inaccurate positioning or waypoint data. Human moral reasoning about the world in these cases is bound to lead to flawed outcomes. A famous example would be the Goalkeeper system used by the British Navy in the Falkland War, which misclassified an incoming Exocet rocket as friendly (i.e. French) when it was, in fact, deployed by the Argentinian enemy. In this case, the system demonstrated its inability to track some morally relevant state of affairs: a rocket being 'friendly' as opposed to just being of the kind usually used by allies; and because of that, it was also unable to track the relevant moral reasons of the human commanders: targeting enemy rockets rather than just targeting rockets with certain material features. More recently, the use of machine learning systems has allegedly led to misclassification of enemy and friendly tanks because the training set had many images of enemy tanks with clouds and many of friendly tanks with cloudless skies.[39] These systems were also tracking irrelevant properties in the training set (i.e. clouds).

The importance of tracking for human control may be further clarified by looking at the following example. A machine learning algorithm learns in a supervised learning process to distinguish between photos of wolves and photos of huskies. After this training it unfailingly classifies new pictures as either huskies or wolves. It turns out, however, that the system just looks at the presence of snow in the background since the majority of wolves' pictures had snow in the background, whereas the husky photos did not. The system was giving good results by tracking the wrong property. If it were presented with a photo of a husky in the snow, it would have classified it as a wolf since it was tracking the colour and texture of the background, not the features of the animal itself.[40]

To avoid developing and deploying systems that are prone to such mistakes, following Fischer and Ravizza conditions of reason-responsiveness, I propose a first necessary condition of Meaningful Human Control:

> **First necessary condition of Meaningful Human Control ('tracking')**: to be under Meaningful Human Control, an AI system should demonstrably and verifiably be *responsive* to the *human* moral reasons relevant to the circumstances – no matter how many system levels, models, software, or devices of whatever nature separate the human being from the ultimate effects in the world. That is, AI systems should *track* (relevant) human moral reasons.[41]

The first condition of Meaningful Human Control is thus cast in terms of a so-called *tracking* relation between human moral capacities to respond to relevant moral reasons and systems' actions. Here, tracking identifies a more robust relation than simple alignment. It requires that not only the states of a system reflect or align with some state of affairs under some given circumstances but the system should also be able to respond in a sufficiently broad range of possible other factual and moral circumstances and to adapt to changes in those circumstances.

The tracking condition does not necessarily imply that those forms of AI and machine learning tools that are probabilistic in nature would not be, per se, eligible candidates for Meaningful Human Control. It just means that whatever its nature or functioning, the system should be able to respond to the world's features in a satisfactory way; this also means that, in practice, we may need to set a reasonable threshold for judging how much reliability in the system responding to the relevant features of the world qualifies as 'tracking' for a particular purpose.

This definition of tracking does not specify who the agents are whose reasons should be tracked in order for the system to be under Meaningful Human Control. The only constraint explicitly contained by the definition is that these should be *human* agents. This means two things: first, a system may be under Meaningful Human Control even if it does not track the reasons of the operator or the deployer under all circumstances, provided it sufficiently tracks the relevant reasons of some other human agents along the chain, such as designers, programmers, legislators, or policymakers. Second, a system may be under Meaningful Human Control and still pursue bad or wrong goals or values. Meaningful Human Control is a necessary but not sufficient condition for a system to be morally or societally good. In Section 8.7, I acknowledge the possibility of a military system that satisfies the tracking condition while at the same time being used for illegal purposes. Indeed, Meaningful Human Control leaves people the possibility to do wrong and take responsibility for that.

However, this does not mean that the tracking condition is morally neutral. In fact, as it requires the system to respond to the human moral reasons *relevant* in the circumstances, it contains an important normative element; establishing whose and which moral reasons are relevant in given circumstances means establishing which normative principle, norms, and values a given system is supposed to follow or reflect. So, even by agreeing that tracking is a necessary condition for Meaningful Human Control, it is still possible to disagree on whether tracking is realised under specific circumstances (due to a normative disagreement on the norms and values to be complied with or realised by a system under specific circumstances).

8.6.2 Tracing

Fischer and Ravizza's second condition for guidance control, the *ownership* condition, may be characterised in terms of a *tracing* condition. The idea of tracing is often used by moral responsibility theorists, and according to US philosopher Manuel Vargas, tracing 'is one of a few things to which nearly all parties in the debate about free will [and moral responsibility, in the anglophone philosophy] appeal to with equal enthusiasm.'[42] The concept of tracing tries to capture the basic intuition that a human agent may be responsible for an outcome even if they do not satisfy the conditions for responsibility *in situ* at the time of the action, provided they were responsible at an earlier time for finding themselves later in the position of not satisfying those conditions. A typical example would be the drunk driver who causes a serious accident while in a state of mental incapacitation – and thus not satisfying the condition for responsibility at that time – but who is responsible for choosing to drink at an earlier moment while knowing that they would drive and that their drunk driving could cause a serious accident. We, therefore, say that the driver's responsibility for causing the accident *traces* back to the moment of their choice to drink (and/or to drive while drunk). Likewise, the driver's responsibility for causing the accident could, in another scenario, be eliminated by tracing back to the moment that someone else tampered with the brakes of their car or, unbeknownst to them, put a drug in their tea.

Moral responsibility theorists are mainly concerned with scenarios where (a) only one individual agent is considered and (b) tracing applies to relations between different actions and mental states of one and the same human agent; in the drunk driving example, between the agent's dangerous driving and their drinking before getting into the car. The challenge for an account of Meaningful Human Control over AI systems is twofold. We need to extend the tracing condition to scenarios where (a) there is more than one human agent, and (b) there are intelligent non-human (sub)systems involved in realising the outcome. This challenge is not completely new. Modern aeroplanes have complex semi-autonomous systems onboard, including collision avoidance systems (CAS), that get them out of harm's way and allow them to rapidly coordinate with other aeroplanes that are dangerously close by. The default policy is to defer to CAS in split-second emergency situations. The pilot can, however, override the system. There are policies in place for the warranted over-ruling of CAS; for example, when the pilot of the other plane that is on a collision course obviously overrides CAS. The pilot is supposed to check their work environment before they enter it and is supposed to establish whether it will allow them to do what they ought to do. Or to have it inspected by others on their behalf. If they fail to do so when they can and find themselves later 'locked-in' in a system's environment that doesn't allow them to discharge their obligation to avoid a collision, then they may be held morally responsible for a negative outcome as their responsibility can be traced back to their failing at an earlier moment to do a proper check.[43]

However, in doing their checks, the pilot is relying on numerous others who have shouldered a part of that burden in the design and production history of the system they are operating. It is unavoidable that the pilot makes certain assumptions and relies on the expertise and goodwill of these engineers, inspectors, mechanics, administrative staff, etc. This means that sometimes the pilot may not be responsible for an accident: it may well be the case that specific others in the chain of system design, production, and maintenance have not discharged their so-called 'second-order responsibility' for the first-order responsibility of the operator *in situ*. If the conditions for first-order operator or user moral responsibility are not satisfied with no fault of the operator, then the responsibility for the accident may be traced back to other agents upstream, even up to the designers of the system, who, for example, may have failed to exemplify a proper understanding of the system.

Based on the analysis of this section, I thus propose to define a second condition for Meaningful Human Control over AI systems in terms of a *tracing relation* between the decision-making system and the technical and moral understanding of some relevant humans involved in the design and deployment of the system. I rely on the general idea of tracing and on Fischer and Ravizza's second condition for guidance control, the *ownership* condition, which requires the agent to properly understand and endorse the mechanism of moral decision-making leading to her action.

Second necessary condition of Meaningful Human Control ('tracing'): For a system to be under Meaningful Human Control, its actions/states should be traceable to a proper moral understanding on the part of the relevant human person(s) who designed or interact with the system, meaning that there is at least one – but ideally more – human agent in the design history or use context involved in designing, programming, operating, and deploying the AI system who understands or is in the position to understand (a) the capabilities of the system and the possible effects in the world of its use and (b) that others may have legitimate moral reactions towards them because of how the system affects the world and the role they play in this chain of events.

Systems whose actions and states are not traceable to a relevant understanding and endorsing by some human person – a designer, a controller, a user, etc. – no matter how intelligent and reason-responsive they may be, are not under Meaningful Human Control. They would be like human actions carried out under psychological manipulation, subliminal persuasion, brainwashing, and indoctrination; here, the system's behaviour is clearly responsive to *someone's* reasons but not to the relevant agent's reasons.

8.7 Application to autonomous weapon systems

To illustrate the tracking and tracing conditions for Meaningful Human Control, let us look at the ethical debate on autonomous weapon systems. In general, critics of autonomous weapon systems seem to be right in stressing that current robotic systems are not able to honour the dual-tracking relation.[44] First, they are likely to fail in tracking the relevant reasons of the human military personnel behind them; in particular, they cannot track the reasoning required by international law, including being guided by considerations of necessity, discrimination, and proportionality. Second, they are not as flexible as to properly adjust their behaviour to the many morally relevant features of the environment in which they operate. If they would be able to distinguish between civilians and combatants, they also need to be able to distinguish between civilians and *civilians apparently involved in armed resistance*, and do so in unstructured and dynamic environments like the ones of present-day battlefields. Therefore, if AI systems were given the possibility to make the decision to initiate an attack without human supervision in an unstructured environment, that would *not* be under Meaningful Human Control.

Some, like the US philosopher Peter Asaro, have argued that the presence of a human operator who takes a 'near-time decision' to initiate each individual attack is a necessary condition to achieve Meaningful Human Control over autonomous weapon systems.[45] Others, like the Dutch military commander Mark Roorda, have suggested that autonomous weapon systems may remain under sufficient control even if no human operator is involved in the 'engagement' (attack) stage of the military operation, provided military commanders in charge of the decision to deploy these weapon systems have followed an appropriate targeting procedure, like, for instance, those of NATO.[46] If these targeting procedures are properly followed, autonomous weapon systems will be (lawfully) used only in circumstances where they can behave according to the human commanders' intentions and reasons. If they are deployed outside these circumstances, the moral and legal responsibility for their behaviour will clearly be traceable to the conscious and culpable decision of the commander. In both cases, the behaviour of the autonomous system will be under Meaningful Human Control, and there won't be any accountability gap.

This argument rightly points to one key general aspect of Meaningful Human Control as analysed in this chapter: being in control does not necessarily require the act of direct controlling from a position that is contiguous in space and time or is a proximate cause, as control in a morally relevant sense allows for technological mediation and separation of the human agent and the relevant moral effects of the acts that they are involved in. Moreover, I concur with the general idea that social and legal practices also contribute to creating people's moral identity and make them legitimate targets of moral and legal responsibility attributions. In fact, whether an agent can legitimately be seen as in control of a certain outcome and thus be legitimately held responsible for that outcome in retrospect also depends on the normative position occupied by that person within a recognised social or legal architecture of duties and responsibilities. In this case, being a commander in a complex military institution. In this perspective, the presence of an appropriate social and legal system of rules is also a necessary component of a socio-technical system that prevents responsibility gaps and achieves Meaningful Human Control over autonomous weapon systems. However, this argument is unsatisfactory as it assumes – as sometimes does – that *current* social and legal practices will be enough to maintain any (future) autonomous weapon system under Meaningful Human Control.

To see why this assumption is unwarranted, it is helpful to look back at the tracing condition for Meaningful Human Control: every action of a decision-making system should be traceable to a proper technical *and* moral understanding on the part of at least one human among those who design and deploy the system, meaning that at least one human agent – but ideally more – is at the same time in the position to understand (a) the capabilities of the system and the possible effects in the world of its use *and* (b) that others may have legitimate moral reactions towards them because of how the system affects the world. It is doubtful whether, under current socio-technical circumstances, military commanders may comply with any of the two sub-conditions. As for (a), the former prosecutor at the International Criminal Tribunal for the former Yugoslavia and legal scholar Dan Saxon remarked that the introduction of AI systems is likely to significantly 'increase the demands on the General's already taxed mental capacity.'[47] He describes how, prior to any deployment of such technology, the general in charge of the action must consider: (1) the range, accuracy, and explosive power of the autonomous weaponry to be directed at the enemy vis-à-vis human-operated weaponry; and the possible presence of civilians in the area; (2) the autonomous weaponry's ability to comply with International Humanitarian Law in the particular battlespace; (3) whether the mission or the expected circumstances of the battlespace may require the exercise of increased levels of human supervision and control over the AI-driven weapon system; (4) whether the general and/or their staff will have the capacity to deactivate the system immediately should conditions require it; (5) the robustness of the software that operates the AI of the system, in particular whether enemy forces may have the ability to tamper with and/or take control over the system(s); and finally (6) the level of training – technical, operational, and with respect to the laws of war – of the human 'operators' or monitors of the AI-driven weapon systems, if any.[48] Given the complexity and difficulty of this task, we cannot assume that just because they are required to do so by the official procedures, commanders will, in practice, be able to properly assess all these variables. For the system to remain under Meaningful Human Control, we need to ensure that military commanders have a sound understanding of the functionalities, capabilities, and limitations of the AI-driven weapon systems available to them.[49] We also must be reminded that the military advantage provided by increasing speed in the acquisition and transmission of information, and in reacting on this information, may influence decisions about acceptable levels of human judgement and permissible levels of autonomy.[50] Finally, the 'automation bias' is a well-studied phenomenon in human–machine interaction more generally: under some circumstances, people tend to overestimate the capabilities of automated systems and, therefore, over-delegate their tasks to them.[51]

Similar considerations apply to condition (b) above. Commanders should understand that others may have legitimate moral reactions towards them because of how the system affects the world; that is, they should realise that they are responsible for what the system does. It has been argued that the tendency for human beings to increasingly depend on computer systems for their decision-making can lead to a reduced sense of responsibility for the consequences of those decisions.[52] In other words, no matter how strict the legal obligation imposed on commanders to take responsibility for the behaviour of autonomous systems, these systems may not be under Meaningful Human Control if commanders do not perceive sufficient ownership of these actions due to a lack of training and experience with the use of these systems.

So, tracing, as defined above, is not only meant, as it were, to help find someone to blame after an accident has occurred; tracing is much more than that. It is an essential component of Meaningful Human Control over a system because it requires that there are always individual humans along the chain who are capable and motivated to take active steps to prevent unwanted outcomes from occurring in the first place.

Two further implications of the above analysis are the following. First, systems that are generally unable to track some relevant moral reasons may still be under Meaningful Human Control in a morally relevant sense if they track the relevant moral reasons of the relevant agents who deploy them. This means that Meaningful Human Control is not a sufficient condition for the morally appropriate behaviour of robotic systems. Consider, for instance, an autonomous weapon system that is unable to comply with the Law of Armed Conflict and is used to perform an unlawful attack by a human operator who is perfectly aware of this inability but decides to use the system anyway to gain military advantage. In this case, I argue that not only the tracing but also the tracking condition is satisfied because though the system is clearly not responsive to some important moral reasons and is not responsive to many relevant features of the environment in which it operates, it is still responsive to the (wrong) relevant moral reasons of the relevant agent who deploys it and it is sensitive to those features of the environment which that agent wants it to be responsive to, in order to achieve their strategic goals. In fact, this would arguably be an unlawful but deliberate attack for which the military commander would clearly be morally and legally culpable.[53] As such, Meaningful Human Control is a necessary but not sufficient condition for the ethical design and use of AI.

Second, autonomous weapon systems may be outside of Meaningful Human Control even if their capabilities and limits are well-known to their human creator and there is no intention to misuse them on the part of the military personnel deploying them. The system may still be out of Meaningful Human Control if there is no individual human agent who is in the position to appreciate the limits in the capabilities of the machine while *at the same time* being aware that the machine's behaviour will be attributed to them; for instance, if the programmers are vividly aware of the limited capacities of the machines but do not feel responsible for their use, because they assume that military commanders will be able to discharge their duty to take the morally relevant decisions about the use of the system; and the military commanders are aware that it is their responsibility to take strategic decisions, but at the same time overestimate the capacity of the systems, due to insufficient training or experience in their use. Here, there is arguably a responsibility gap as no individual agent satisfies the tracing condition.

8.8 Bridging responsibility gaps with Meaningful Human Control

This section explains how Meaningful Human Control may help preventing responsibility gaps. Its main points are summarised in Table 8.1 below. In the previous chapter, I rejected what I called 'technical solutionism,' the idea that technology can become more controllable by just improving its technical features. The Meaningful Human Control approach recognises that human control requires not only new technical features, for instance, to promote technical transparency (e.g. explainable AI) but also different social, institutional, and legal arrangements to promote organisational and political transparency about the control roles. The tracking condition for Meaningful Human Control demands a systematic specification of the control roles based on the system's 'responsiveness' to human reasons. This requires, as a first step, a map of all the relevant agents involved in the design, control, regulation, and use of the system, as well as the extent to which their reasons, values, and intentions are (or should be) reflected in its behaviour.[54]

As a further illustration, consider again the 2018 Uber accident,[55] in which a car in automated mode but with a human operator onboard ran over a pedestrian due to a technical failure in the sensing system and lack of sufficient oversight from the on-board operator. Who was in control of the system and thus responsible for the accident? The legal liability process clearly relied on a narrow notion of control: the onboard operator was deemed in control of the system as the vehicle was responsive to her behaviour. She was in the position to intervene and avoid the crash, and since she failed to behave as requested by her role, she was considered negligent *and* liable for the death.[56] This answer sounds too simplistic: more 'controllers' are involved in this story. By using Mecacci and Santoni de Sio's (2020) analysis of the tracking condition, we can more easily identify other actors whose higher-level plans and goals played or could have played a control role in the story and may, therefore, have some form of responsibility in it. Just to mention two: the vehicle's dangerous behaviour was clearly also dependent on the explicit intention of the owner company (Uber) to test new not-yet-safe self-driving features on public roads with the intention of pushing their technical development, and the opportunity to do these tests was created by Arizona's authorities with the goal of attracting (big) tech companies in their State.[57] So it seems fair to attribute some *prima facie* moral culpability for the death of the pedestrian also to them. This, in turn, triggers a moral duty for them to explain to the victim's relatives and to the broader public the reasons for their strategic and political choices (i.e. moral and public accountability). Correspondingly, the map offers a starting point for designers, policymakers, and researchers to identify the different active responsibilities of the agents involved, as well as a general schema to identify the policy, legal, technical, and organisational interventions needed to ensure that the system is sufficiently responsive to the right reasons of the relevant agents.

Meaningful Human Control requires that socio-technical systems are also systematically designed to avoid gaps not only in legal liability but also, following the distinction introduced in the previous chapter, in moral culpability, accountability and active responsibility. The tracing condition proposes that a system can remain under Meaningful Human Control only in the presence of a solid alignment between the system and the technical, motivational and moral capacities of the relevant agents involved, with different roles, in the design, control and use of the system. The direct goal of this condition is promoting a fair distribution of moral culpability, thereby avoiding two undesired results. First, it avoids scapegoating, i.e. agents being held culpable without having a fair capacity to avoid wrongdoing.[58] In the case of automated driving systems, the driver could lack relevant technical and motivational capacities due to insufficient training. Second, it avoids impunity for avoidable accidents, i.e. culpability gaps, where it is impossible to legitimately blame anybody as no individual agent possesses all the relevant capacities. For example, the managers/designers could have the technical capacity but not the moral motivation to avoid accidents and the drivers could have the moral motivation but not the skills to prevent accidents.

The tracing condition also helps address accountability and active responsibility gaps. If a person or organisation should be morally or publicly accountable, then they must also possess the specific capacity to discharge this duty: if, for instance, a doctor has to remain accountable to their patients for their decisions, then they should maintain the capacity and motivation to understand the functioning of the AI system they use and to explain their decision to the patients. And if a public organisation has to remain accountable to their citizens, then arguably public officers and civil servants, not IT experts and private companies, should maintain the capacity and motivation to understand and explain the behaviour of their organisation to the citizens, no matter how much technology is involved in the decision. So, the tracing condition poses a strong accent on the individual, context-relative capacities needed for the relevant human actors to remain accountable to different subjects and in relation to different activities.

Table 8.1 Conditions for Meaningful Human Control and how they prevent responsibility gaps

Meaningful Human Control		
Conditions	*Operationalisation*	*Payoffs for different types of responsibility*
Tracking condition: The system is responsive to the relevant reasons of relevant human agents	Mapping the agents involved in the system, their reasons or intentions, and their relations with the systems	*Culpability and accountability:* Extends culpability and accountability attribution to a wider range of agents (e.g. to companies and regulators in case of self-driving car accidents)
		Active responsibility: Supports managers, designers, policymakers, and researchers in identifying potential value tensions and recognising their own responsibility to design for all the relevant reasons
Tracing condition: Alignment between the system and human capacities (technical, motivational, and moral)	Analysing the capacities of the human agents in the system and checking that the system sufficiently reflects them	*Culpability:* Allows for fairer attributions, thus avoiding both scapegoating and impunity for accidents (e.g. blaming the driver or nobody for an avoidable self-driving car accident)
		Moral accountability: Requires that the relevant agents understand the AI working processes and their role in the decision-making process (e.g. a medical doctor using AI for diagnosis)
		Public accountability: Requires that public officers (not only IT experts and private companies) remain able and motivated to understand the AI working processes and their duty to respond to the public for it
		Active responsibility: Supports various actors in developing the knowledge, capacity, opportunity, and motivation to discharge their various responsibilities in relation to AI

The analysis and promotion of (moral) capacities is also key to ensuring the discharge of various active responsibilities by various actors. The Responsible Innovation approach invites stakeholders to be responsive to one another and to the relevant values in the process of technological development.[59] The tracing condition also demands that these stakeholders are effectively supported in acquiring the context-specific capacity, opportunity, and motivation to do so. For instance, across the years, the aviation sector has built up a culture of safety and the medical profession a culture of care; as also recently proposed by the EU expert group report on ethical issues with driverless mobility[60] a new 'culture of responsibility' – new practices, incentives, and identities – should be promoted within private and public organisations in relation to the design and use of AI-based systems.

Notes

1 Jürgen Altmann et al., 'Armed Military Robots: Editorial,' *Ethics and Information Technology* 15, no. 2 (20 June 2013): 73, https://doi.org/10.1007/s10676-013-9318-1.

2 John Markoff, 'Fearing Bombs That Can Pick Whom to Kill,' *The New York Times*, 12 November 2014.

3 Simon Parkin, 'Killer Robots: The Soldiers That Never Sleep,' 16 July 2015, https://www.bbc.com/future/article/20150715-killer-robots-the-soldiers-that-never-sleep.

4 Robyn Dixon, 'Azerbaijan's Drones Owned the Battlefield in Nagorno-Karabakh — and Showed Future of Warfare,' *Washington Post*, 12 November 2020, https://www.washingtonpost.com/world/europe/nagorno-karabkah-drones-azerbaijan-aremenia/2020/11/11/441bcbd2-193d-11eb-8bda-814ca56e138b_story.html.

5 James Vincent, 'Have Autonomous Robots Started Killing in War? The Reality Is Messier than It Appears,' 2022, https://www.theverge.com/2021/6/3/22462840/killer-robot-autonomous-drone-attack-libya-un-report-context.

6 David Hambling, 'Ukrainian AI Attack Drones May Be Killing without Human Oversight,' *New Scientist*, 13 October 2023, https://www.newscientist.com/article/2397389-ukrainian-ai-attack-drones-may-be-killing-without-human-oversight/#.

7 Robin Fontes and Jorrit Kamminga, 'Ukraine a Living Lab for AI Warfare,' *National Defense*, 24 March 2023, https://www.nationaldefensemagazine.org/articles/2023/3/24/ukraine-a-living-lab-for-ai-warfare.

8 Harry Davies, Bethan McKernan, and Dan Sabbagh, '"The Gospel": How Israel Uses AI to Select Bombing Targets in Gaza,' *The Guardian*, 1 December 2023, https://www.theguardian.com/world/2023/dec/01/the-gospel-how-israel-uses-ai-to-select-bombing-targets?CMP=share_btn_tw.

9 Bonnie Docherty, 'Losing Humanity. The Case against Killer Robots,' *Human Rights Watch* (Human Rights Watch and Harvard International Human Rights Clinic, November 2012), https://www.hrw.org/report/2012/11/19/losing-humanity/case-against-killer-robots.

10 Merel A. C. Ekelhof, 'Complications of a Common Language: Why It Is so Hard to Talk about Autonomous Weapons,' *Journal of Conflict and Security Law* 22, no. 2 (1 July 2017): 311–31, https://doi.org/10.1093/jcsl/krw029.
11 Article 36, 'Autonomous Weapons, Meaningful Human Control and the CCW,' (Article 36, 2014), http://www.article36.org/weapons-review/autonomous-weapons-meaningful-human-control-and-the-ccw/.
12 Sarah Knuckey, 'Governments Conclude First (Ever) Debate on Autonomous Weapons: What Happened and What's Next,' 2014, https://www.justsecurity.org/10518/autonomous-weapons-intergovernmental-meeting/; Michael C. Horowitz and Paul Scharre, 'Meaningful Human Control in Weapon Systems: A Primer,' (Center for a New American Security, 2015), https://www.cnas.org/publications/reports/meaningful-human-control-in-weapon-systems-a-primer.
13 Horowitz and Scharre, 'Meaningful Human Control in Weapon Systems: A Primer.'
14 Kerstin Vignard, 'The Weaponization of Increasingly Autonomous Technologies: Considering How Meaningful Human Control Might Move Discussion Forward,' *UNIDIR Resources*, no. 2 (2014): 3.
15 Stefan Buijsman et al., 'Responsible AI in Military Applications,' *Ethics and Information Technology*, 2023, https://link.springer.com/collections/cghhijhhji.
16 Merel Ekelhof, 'Moving Beyond Semantics on Autonomous Weapons: Meaningful Human Control in Operation,' *Global Policy* 10, no. 3 (September 2019): 343–48, https://doi.org/10.1111/1758-5899.12665; Daniele Amoroso and Guglielmo Tamburrini, 'What Makes Human Control Over Weapon Systems 'Meaningful'?,' *ICRAC Working Paper Series #4*, 2019; Article 36, 'Autonomous Weapons, Meaningful Human Control and the CCW'; Heather M. Roff and Richard Moyes, 'Meaningful Human Control, Artificial Intelligence and Autonomous Weapons,' 2016; Horowitz and Scharre, 'Meaningful Human Control in Weapon Systems: A Primer.'
17 Horowitz and Scharre, 'Meaningful Human Control in Weapon Systems: A Primer.'
18 Dan Saxon, 'Autonomous Drones and Individual Criminal Responsibility,' in *Drones and Responsibility: Legal, Philosophical, and Socio-Technical Perspectives on the Use of Remotely Controlled Weapons*, eds. Ezio Di Nucci and Filippo Santoni de Sio (London and New York: Routledge, 2016), 17–46.
19 Together with a multidisciplinary group of philosophers, lawyers, and engineers, I have recently prepared a handbook that offers a comprehensive exploration of the topic of Meaningful Human Control from different disciplines and applied to different domains, see Giulio Mecacci et al., *Research Hanbook of Meaningful Human Control over AI Systems* (Cheltenham: Edward Elgar, forthcoming).
20 Simeon C. Calvert et al., 'Gaps in the Control of Automated Vehicles on Roads,' *IEEE Intelligent Transportation Systems Magazine*, 2020, 1, https://doi.org/10.1109/MITS.2019.2926278; Simeon C. Calvert et al., 'A Human Centric Framework for the Analysis of Automated Driving Systems Based on Meaningful Human Control,' *Theoretical Issues in Ergonomics Science*, 10 December 2019, 1–29, https://doi.org/10.1080/1463922X.2019.1697390; Simeon C. Calvert et al., 'Full Platoon Control in Truck Platooning: A Meaningful Human Control Perspective,' in *21st International Conference on Intelligent Transportation Systems (ITSC)* (Maui, HI: IEEE, 2018), 3320–26, https://doi.org/10.1109/ITSC.2018.8570013; Daniël D. Heikoop et al., 'Human Behaviour with Automated Driving Systems: A Quantitative Framework for Meaningful Human Control,' *Theoretical Issues in Ergonomics Science* 20, no. 6 (2 November 2019): 711–30, https://doi.org/10.1080/1463922X.2019.1574931; Giulio Mecacci and Filippo

Santoni de Sio, 'Meaningful Human Control as Reason-Responsiveness: The Case of Dual-Mode Vehicles,' *Ethics and Information Technology* 22, no. 2 (19 June 2020): 103–15, https://doi.org/10.1007/s10676-019-09519-w.

21 Eva Maria Hille, Patrik Hummel, and Matthias Braun, 'Meaningful Human Control over AI for Health? A Review,' *Journal of Medical Ethics*, 20 September 2023, https://doi.org/10.1136/jme-2023-109095.

22 Melissa Heikkilä, 'Dutch Scandal Serves as a Warning for Europe over Risks of Using Algorithms,' *POLITICO* (blog), 29 March 2022, https://www.politico.eu/article/dutch-scandal-serves-as-a-warning-for-europe-over-risks-of-using-algorithms/.

23 Article 29 (Working Party), 'Guidelines on Automated Individual Decision-Making and Profiling for the Purposes of Regulation 2016/679 (2018) 21,' (2018); Ben Wagner, 'Liable, but Not in Control? Ensuring Meaningful Human Agency in Automated Decision-Making Systems,' *Policy & Internet* 11, no. 1 (March 2019): 104–22, https://doi.org/10.1002/poi3.198.

24 This chapter is based on a previously published article co-authored with Jeroen van den Hoven. Credit is therefore due to Jeroen van den Hoven for the content of this chapter. This article offers a more detailed discussion of the theory of Meaningful Human Control. Filippo Santoni de Sio and Jeroen van den Hoven, 'Meaningful Human Control over Autonomous Systems: A Philosophical Account,' *Frontiers in Robotics and AI* 5 (February 2018): 15, https://doi.org/10.3389/frobt.2018.00015. Since 2018, I have further developed the theory of Meaningful Human Control as reason-responsiveness with Giulio Mecacci and other colleagues. This is why in this chapter I use the pronouns 'we'/'our' to refer to the theory.

25 John Martin Fischer and Mark Ravizza, *Responsibility and Control : A Theory of Moral Responsibility* (Cambridge: Cambridge University Press, 1998).

26 Batya Friedman and Peter H. Kahn Jr, 'Human Values, Ethics, and Design,' in *The Human-Computer Interaction Handbook*, eds. Andrew Sears and Julie A. Jacko (Boca Raton, Florida: CRC Press, 2009), 1277; Janet Davis and Lisa P. Nathan, 'Value Sensitive Design: Applications, Adaptations, and Critiques,' in *Handbook of Ethics, Values, and Technological Design: Sources, Theory, Values and Application Domains*, eds. Jeroen van den Hoven, Pieter E. Vermaas, and Ibo van de Poel (Dordrecht: Springer, 2015), 11–40, https://doi.org/10.1007/978-94-007-6970-0_3.

27 Jeroen van den Hoven, 'Value Sensitive Design and Responsible Innovation,' in *Responsible Innovation: Managing the Responsible Emergence of Science and Innovation in Society*, eds. Richard Owen, John Bessant, and Maggy Heintz (Hoboken, NJ: John Wiley & Sons, Ltd, 2013), 75–83, http://onlinelibrary.wiley.com/doi/10.1002/9781118551424.ch4/summary.

28 Robert Kane, *The Significance of Free Will* (New York: Oxford University Press, 1996); Peter Van Inwagen, *An Essay on Free Will* (Oxford: Clarendon Press, 1983); David Hodgson, *Rationality + Consciousness = Free Will* (New York: Oxford University Press, 2012).

29 Skepticism about free will and moral responsibility may be generally grounded in causal determinism as in Derk Pereboom, *Living Without Free Will* (Cambridge: Cambridge University Press, 2001), in the pervasiveness of luck factors as in Neil Levy, *Hard Luck: How Luck Undermines Free Will and Moral Responsibility* (Oxford: Oxford University Press, 2011), in neurobiological reductionism, as in Jonathan D. Cohen and Joshua Greene, 'For the Law, Neuroscience Changes Nothing and Everything,' *Philosophical Transactions of the Royal Society of London. Series B, Biological Sciences* 359 (2004): 1775–85, or in the perva-

siveness of unconscious psychological mechanisms, as in Gregg D. Caruso, *Free Will and Consciousness: A Determinist Account of the Illusion of Free Will* (Plyomouth: Lexington Books, 2012).

30 Harry G. Frankfurt, 'Freedom of the Will and the Concept of a Person,' *The Journal of Philosophy* 68, no. 1 (14 January 1971): 5, https://doi.org/10.2307/2024717; Daniel C. Dennett, *Elbow Room: The Varieties of Free Will Worth Wanting* (Cambridge, MA: The MIT Press, 1984); Fischer and Ravizza, *Responsibility and Control: A Theory of Moral Responsibility*; Daniel C. Dennett, *Freedom Evolves* (London: Penguin, 2004).

31 This idea often supports some deep moral concerns about AI and the specific claim that autonomous weapon systems should not only be avoided due to the risk of negative moral and legal consequences and responsibility gaps, but they are *mala in se* (intrinsically bad) as they violate human dignity by bypassing human decision-making. Peter Asaro, 'On Banning Autonomous Weapon Systems: Human Rights, Automation, and the Dehumanization of Lethal Decision-Making,' *International Review of the Red Cross*, 2012, https://www.icrc.org/eng/resources/documents/article/review-2012/irrc-886-asaro.htm; Peter Asaro, 'Jus Nascendi, Robotic Weapons and the Martens Clause,' in *Robot Law*, eds. Ryan Calo, A. Michael Froomkin, and Ian Kerr (Cheltenham: Edward Elgar Publishing, 2016), 402.

32 Fischer and Ravizza, *Responsibility and Control: A Theory of Moral Responsibility*.

33 Fischer and Ravizza.

34 Fischer and Ravizza, 35–36.

35 Fischer and Ravizza, 207–39; John Martin Fischer, 'Responsibility and Manipulation,' *The Journal of Ethics* 8, no. 2 (2004): 146.

36 Andy Clark and David Chalmers, 'The Extended Mind,' *Analysis* 58, no. 1 (1 January 1998): 7–19, https://doi.org/10.1111/1467-8284.00096; Mark Alfano, *Character as Moral Fiction* (Cambridge: Cambridge University Press, 2013).

37 Ezio Di Nucci and Filippo Santoni de Sio, 'Who's Afraid of Robots? Fear of Automation and the Ideal of Direct Control,' in *Roboethics in Film*, eds. Fiorella Battaglia and Nathalie Weidenfeld (Pisa: Pisa University Press, 2014).

38 The use of the terms 'tracking' (in this section) and 'tracing' (in the next) follows the philosophical usage and does not exactly match the use in the military or in the transport domain. In fact, I use the notion of 'tracking' by following Robert Nozick's account of knowledge as 'truth tracking.' Robert Nozick, *Philosophical Explanations* (Cambridge, MA: Harvard University Press, 1981). For an explanation, see Santoni de Sio and van den Hoven, 'Meaningful Human Control over Autonomous Systems: A Philosophical Account,' 6–7.

39 Eliezer Yudowsky, 'Artificial Intelligence as a Positive and Negative Factor in Global Risk,' in *Global Catastrophic Risks*, eds. Nick Bostrom and Milan M. Ćirković (New York: Oxford University Press, 2008), 308–45.

40 Marco Tulio Ribeiro, Sameer Singh, and Carlos Guestrin, "'Why Should I Trust You?' Explaining the Predictions of Any Classifier,' in *Proceedings of the 22nd ACM SIGKDD International Conference on Knowledge Discovery and Data Mining - KDD '16* (New York: ACM Press, 2016), 1135–44, https://doi.org/10.1145/2939672.2939778.

41 What should count as a reason is a controversial topic in philosophy. Some claim that reasons should be defined normatively, that is, as the facts the may justify an action (normative reasons); for instance, a reason to take the umbrella is the fact that it rains; other claims that also motivational states like intentions should count as reasons (motivating reason): in this sense, a

reason to take the umbrella is also that I believe it is raining and I do not want to get wet. This is a reason to act, whether as a matter of fact it rains or not. Herman Veluwenkamp, 'Reasons for Meaningful Human Control,' *Ethics and Information Technology* 24, no. 4 (23 November 2022): 51, https://doi.org/10.1007/s10676-022-09673-8. Veluwenkamp claims that the theory of Meaningful Human Control should rely only on normative reasons, like arguably does the theory of Fischer and Ravizza. I tend to disagree, and in previous work we have also included motivating reasons in our account of Meaningful Human Control. The reason is (pun not intended) that I believe that a system can be under Meaningful Human Control also if the relevant agents are motivated by the wrong reasons, provided that their wrong motivation is not due to the lack of capacity and opportunities to understand the system and their moral obligations within it, that is provided the tracing condition, presented below, is also satisfied.

42 Kevin Timpe, 'Tracing and the Epistemic Condition on Moral Responsibility,' *The Modern Schoolman* 88, no. 1 (1 April 2011): 5–28, https://doi.org/10.5840/schoolman2011881/22.

43 For a broader discussion, see Jeroen van den Hoven, 'Moral Responsibility, Public Office and Information Technology,' in *Public Administration in an Information Age: A Handbook*, eds. I. Th. M. Snellen and Wim B. H. J. van de Donk (Amsterdam: IOS Press, 1998), 579: Maarten Franssen, 'Design for Values and Operator Roles in Sociotechnical Systems,' in *Handbook of Ethics, Values, and Technological Design: Sources, Theory, Values and Application Domains*, eds. Jeroen van den Hoven, Pieter E. Vermaas, and Ibo van de Poel (Dordrecht: Springer 2015), 118–48, https://doi.org/10.1007/978-94-007-6970-0.

44 Peter Asaro, 'How Just Could a Robot War Be?,' in *Proceedings of the 2008 Conference on Current Issues in Computing and Philosophy*, eds. Philip A. E. Brey, Adam Briggle, and Katinka Waelbers (Amsterdam: Ios Press, 2008), 205; Noel Sharkey, 'Automated Killers and the Computing Profession,' *Computer* 40, no. 11 (November 2007): 122–24, https://doi.org/10.1109/MC.2007.372.

45 Asaro, 'On Banning Autonomous Weapon Systems.'

46 Mark Roorda, 'NATO's Targeting Process: Ensuring Human Control over (and Lawful Use of) "Autonomous" Weapons,' in *Autonomous Systems: Issues for Defence Policymakers*, eds. Andrew P. Williams and Paul D. Scharre (HQ Sact, 2016).

47 Saxon, 'Autonomous Drones and Individual Criminal Responsibility.'

48 Saxon.

49 Saxon.

50 Saxon.

51 Antonio Coco, 'Exploring the Impact of Automation Bias and Complacency on Individual Criminal Responsibility for War Crimes,' *Journal of International Criminal Justice*, 22 September 2023, https://doi.org/10.1093/jicj/mqad034.

52 M. L. Cummings, 'Automation and Accountability in Decision Support System Interface Design,' *Journal of Technology Studies* 32, no. 1 (2006): 23–31; Mark Coeckelbergh, 'Drones, Information Technology, and Distance: Mapping the Moral Epistemology of Remote Fighting,' *Ethics and Information Technology* 15, no. 2 (8 June 2013): 87–98, https://doi.org/10.1007/s10676-013-9313-6; Saxon, 'Autonomous Drones and Individual Criminal Responsibility.'

53 Saxon.

54 Together with Giulio Mecacci I have started developing such a map in relation to automated driving systems (dual-mode vehicles), see Mecacci and Santoni de

Sio, 'Meaningful Human Control as Reason-Responsiveness: The Case of Dual-Mode Vehicles.'

55 Tina Bellon, 'Fatal U.S. Self-Driving Auto Accident Raises Novel Legal Questions,' *Reuters*, 2018, https://www.reuters.com/article/us-autos-selfdriving-uber-liability-anal/fatal-u-s-self-driving-auto-accident-raises-novel-legal-questions-idUSKBN1GW2SP; Jack Stilgoe, 'Who Killed Elaine Herzberg?,' in Jack Stilgoe, *Who's Driving Innovation?* (Cham: Palgrave MacMillan, 2020), 1–6, https://doi.org/10.1007/978-3-030-32320-2_1.

56 Lauren Smiley, 'The Legal Saga of Uber's Fatal Self-Driving Car Crash Is Over,' *Wired*, 28 July 2023, https://www.wired.com/story/ubers-fatal-self-driving-car-crash-saga-over-operator-avoids-prison/.

57 National Transportation Safety Board, 'Collision Between Vehicle Controlled by Developmental Automated Driving System and Pedestrian, Tempe, Arizona 18 March 2018,' Highway Accident Report NTSB/HAR-19/03. Washington, DC, 2019.

58 Madeleine Clare Elish, 'Moral Crumple Zones: Cautionary Tales in Human-Robot Interaction,' *Engaging Science, Technology, and Society* 5 (23 March 2019): 40, https://doi.org/10.17351/ests2019.260.

59 Jack Stilgoe, Richard Owen, and Phil Macnaghten, 'Developing a Framework for Responsible Innovation,' *Research Policy* 42, no. 9 (2013): 1568–80, https://doi.org/10.1016/j.respol.2013.05.008.

60 Jean-François Bonnefon, David Černy, John Danaher, Nathalie Devillier, Veronika Johansson, Tatiana Kovacikova, Marieke Martens, Miloš Mladenovic, Paula Palade, Nick Reed, Filippo Santoni de Sio, Stavroula Tsinorema, Sandra Wachter, and Karolina Zawieska, *Horizon 2020 Commission Expert Group to Advise on Specific Ethical Issues Raised by Driverless Mobility (E03659). Ethics of Connected and Automated Vehicles: Recommendations on Road Safety, Privacy, Fairness, Explainability and Responsibility* (Luxembourg: European Commission, 2020).

References

Alfano, Mark. *Character as Moral Fiction*. Cambridge: Cambridge University Press, 2013.

Altmann, Jürgen, Peter Asaro, Noel Sharkey, and Robert Sparrow. 'Armed Military Robots: Editorial.' *Ethics and Information Technology* 15, no. 2 (20 June 2013): 73–76. https://doi.org/10.1007/s10676-013-9318-1.

Amoroso, Daniele, and Guglielmo Tamburrini. *What Makes Human Control Over Weapon Systems 'Meaningful'?* ICRAC Working Paper Series No. 4, 2019.

Article 29 (Working Party). Guidelines On Automated Individual Decision-Making and Profiling for the Purposes of Regulation 2016/679 (2018) 21 (2018).

Article 36. 'Autonomous Weapons, Meaningful Human Control and the CCW.' Article 36, 2014. http://www.article36.org/weapons-review/autonomous-weapons-meaningful-human-control-and-the-ccw/.

Asaro, Peter. 'How Just Could a Robot War Be?' In *Proceedings of the 2008 Conference on Current Issues in Computing and Philosophy*, Vol. 205, edited by Philip A. E. Brey, Adam Briggle, and Katinka Waelbers. Amsterdam, Ios Press, 2008. https://dl.acm.org/citation.cfm?id=1566243.

———. 'Jus Nascendi, Robotic Weapons and the Martens Clause.' In *Robot Law*, edited by Ryan Calo, A. Michael Froomkin, and Ian Kerr. Cheltenham: Edward Elgar Publishing, 2016.

———. 'On Banning Autonomous Weapon Systems: Human Rights, Automation, and the Dehumanization of Lethal Decision-Making.' *International Review of the Red Cross* (2012). https://www.icrc.org/eng/resources/documents/article/review-2012/irrc-886-asaro.htm.

Bellon, Tina. 'Fatal U.S. Self-Driving Auto Accident Raises Novel Legal Questions.' *Reuters*, 2018. https://www.reuters.com/article/us-autos-selfdriving-uber-liability-anal/fatal-u-s-self-driving-auto-accident-raises-novel-legal-questions-idUSKBN1GW2SP.

Bonnefon, Jean-François, David Černy, John Danaher, Nathalie Devillier, Veronika Johansson, Tatiana Kovacikova, Marieke Martens, Miloš Mladenovic, Paula Palade, Nick Reed, Filippo Santoni de Sio, Stavroula Tsinorema, Sandra Wachter, Karolina Zawieska. *Horizon 2020 Commission Expert Group to Advise on Specific Ethical Issues Raised by Driverless Mobility (E03659). Ethics of Connected and Automated Vehicles: Recommendations on Road Safety, Privacy, Fairness, Explainability and Responsibility*. Luxembourg: European Commission, 2020.

Buijsman, Stefan, Filippo Santoni de Sio, Mignon Hagemeijer, Daniele Amoroso, Jurriaan van Diggelen, Bérénice Boutin, Catholijn Jonker, Ilse Verdiesen, and Jeroen van den Hoven. 'Responsible AI in Military Applications.' *Ethics and Information Technology*, 2023. https://link.springer.com/collections/cghhijhhji.

Calvert, Simeon C., Daniël D. Heikoop, Giulio Mecacci, and Bart van Arem. 'A Human Centric Framework for the Analysis of Automated Driving Systems Based on Meaningful Human Control.' *Theoretical Issues in Ergonomics Science* (10 December 2019): 1–29. https://doi.org/10.1080/1463922X.2019.1697390.

Calvert, Simeon C., Giulio Mecacci, Bart van Arem, Filippo Santoni de Sio, Daniel D. Heikoop, and Marjan Hagenzieker. 'Gaps in the Control of Automated Vehicles on Roads.' *IEEE Intelligent Transportation Systems Magazine* (2020): 1. https://doi.org/10.1109/MITS.2019.2926278.

Calvert, Simeon C., Giulio Mecacci, Daniel D. Heikoop, and Filippo Santoni De Sio. 'Full Platoon Control in Truck Platooning: A Meaningful Human Control Perspective.' *2018 21st International Conference on Intelligent Transportation Systems (ITSC)*, 3320–26. Maui, HI: IEEE, 2018. https://doi.org/10.1109/ITSC.2018.8570013.

Caruso, Gregg D. *Free Will and Consciousness : A Determinist Account of the Illusion of Free Will*. Plyomouth: Lexington Books, 2012.

Clark, Andy, and David Chalmers. 'The Extended Mind.' *Analysis* 58, no. 1 (1 January 1998): 7–19. https://doi.org/10.1111/1467-8284.00096.

Coco, Antonio. 'Exploring the Impact of Automation Bias and Complacency on Individual Criminal Responsibility for War Crimes.' *Journal of International Criminal Justice* (22 September 2023). https://doi.org/10.1093/jicj/mqad034.

Coeckelbergh, Mark. 'Drones, Information Technology, and Distance: Mapping the Moral Epistemology of Remote Fighting.' *Ethics and Information Technology* 15, no. 2 (8 June 2013): 87–98. https://doi.org/10.1007/s10676-013-9313-6.

Cohen, Jonathan D., and Joshua Greene. 'For the Law, Neuroscience Changes Nothing and Everything.' *Philosophical Transactions of the Royal Society of London. Series B, Biological Sciences* 359 (2004): 1775–85.

Cummings, M. L. 'Automation and Accountability in Decision Support System Interface Design.' *Journal of Technology Studies* 32, no. 1 (2006): 23–31.

Davis, Janet, and Lisa P. Nathan. 'Value Sensitive Design: Applications, Adaptations, and Critiques.' In *Handbook of Ethics, Values, and Technological Design: Sources, Theory, Values and Application Domains*, edited by Jeroen van den Hoven, Pieter E. Vermaas, and Ibo van de Poel, 11–40. Dordrecht: Springer, 2015. https://doi.org/10.1007/978-94-007-6970-0_3.

Dennett, Daniel C. *Elbow Room: The Varieties of Free Will Worth Wanting*. Cambridge, MA: The MIT Press, 1984.

———. *Freedom Evolves*. New York: Penguin, 2004.

Di Nucci, Ezio, and Filippo Santoni de Sio. 'Who's Afraid of Robots? Fear of Automation and the Ideal of Direct Control.' In *Roboethics in Film*, edited by Fiorella Battaglia and Nathalie Weidenfeld. Pisa: Pisa University Press, 2014.

Dixon, Robyn. 'Azerbaijan's Drones Owned the Battlefield in Nagorno-Karabakh — And Showed Future of Warfare.' *Washington Post*, 12 November 2020. https://www.washingtonpost.com/world/europe/nagorno-karabkah-drones-azerbaijan-aremenia/2020/11/11/441bcbd2-193d-11eb-8bda-814ca56e138b_story.html.

Docherty, Bonnie. *Losing Humanity. The Case against Killer Robots*. Human Rights Watch and Harvard International Human Rights Clinic, November 2012. https://www.hrw.org/report/2012/11/19/losing-humanity/case-against-killer-robots.

Ekelhof, Merel A. C. 'Complications of a Common Language: Why It Is so Hard to Talk about Autonomous Weapons.' *Journal of Conflict and Security Law* 22, no. 2 (1 July 2017): 311–31. https://doi.org/10.1093/jcsl/krw029.

———. 'Moving Beyond Semantics on Autonomous Weapons: Meaningful Human Control in Operation.' *Global Policy* 10, no. 3 (September 2019): 343–48. https://doi.org/10.1111/1758-5899.12665.

Elish, Madeleine Clare. 'Moral Crumple Zones: Cautionary Tales in Human-Robot Interaction.' *Engaging Science, Technology, and Society* 5 (23 March 2019): 40. https://doi.org/10.17351/ests2019.260.

Fischer, John Martin. 'Responsibility and Manipulation.' *The Journal of Ethics* 8, no. 2 (2004): 145–77.

Fischer, John Martin, and Mark Ravizza. *Responsibility and Control : A Theory of Moral Responsibility*. Cambridge: Cambridge University Press, 1998.

Fontes, Robin, and Jorrit Kamminga. 'Ukraine A Living Lab for AI Warfare.' *National Defense*, 24 March 2023. https://www.nationaldefensemagazine.org/articles/2023/3/24/ukraine-a-living-lab-for-ai-warfare.

Frankfurt, Harry G. 'Freedom of the Will and the Concept of a Person.' *The Journal of Philosophy* 68, no. 1 (14 January 1971): 5. https://doi.org/10.2307/2024717.

Franssen, Maarten. 'Design for Values and Operator Roles in Sociotechnical Systems.' In *Handbook of Ethics, Values, and Technological Design: Sources, Theory, Values and Application Domains*, edited by Jeroen van den Hoven,

Pieter E. Vermaas, and Ibo van de Poel, 118–48. Dordrecht: Springer, 2015. https://doi.org/10.1007/978-94-007-6970-0.

Friedman Batya, and Peter H. Jr. Kahn. 'Human Values, Ethics, and Design.' In *The Human-Computer Interaction Handbook*, edited by Andrew Sears and Julie A. Jacko, 1277. Boca Raton, FL: CRC Press, 2009.

Hambling, David. 'Ukrainian AI Attack Drones May Be Killing without Human Oversight.' New Scientist, 13 October 2023. https://www.newscientist.com/article/2397389-ukrainian-ai-attack-drones-may-be-killing-without-human-oversight/#.

Heikkilä, Melissa. 'Dutch Scandal Serves as a Warning for Europe over Risks of Using Algorithms.' *Politico* (blog), 29 March 2022. https://www.politico.eu/article/dutch-scandal-serves-as-a-warning-for-europe-over-risks-of-using-algorithms/.

Heikoop, Daniël D., Marjan Hagenzieker, Giulio Mecacci, Simeon Calvert, Filippo Santoni De Sio, and Bart Van Arem. 'Human Behaviour with Automated Driving Systems: A Quantitative Framework for Meaningful Human Control.' *Theoretical Issues in Ergonomics Science* 20, no. 6 (2 November 2019): 711–30. https://doi.org/10.1080/1463922X.2019.1574931.

Hille, Eva Maria, Patrik Hummel, and Matthias Braun. 'Meaningful Human Control Over AI for Health? A Review.' *Journal of Medical Ethics* (20 September 2023). https://doi.org/10.1136/jme-2023-109095.

Hodgson, David. *Rationality + Consciousness = Free Will*. New York: Oxford University Press, 2012. https://doi.org/10.1093/acprof:oso/9780199845309.001.0001.

Horowitz, Michael C., and Paul Scharre. *Meaningful Human Control in Weapon Systems: A Primer*. Center for a New American Security, 2015. https://www.cnas.org/publications/reports/meaningful-human-control-in-weapon-systems-a-primer.

Kane, Robert. *The Significance of Free Will*. New York: Oxford University Press, 1996.

Knuckey, Sarah. 'Governments Conclude First (Ever) Debate on Autonomous Weapons: What Happened and What's Next.' 2014. https://www.justsecurity.org/10518/autonomous-weapons-intergovernmental-meeting/.

Levy, Neil. *Hard Luck: How Luck Undermines Free Will and Moral Responsibility*. Oxford: Oxford University Press, 2011.

Markoff, John. 'Fearing Bombs That Can Pick Whom to Kill.' *The New York Times*, 12 November 2014.

Mecacci, Giulio, and Filippo Santoni de Sio. 'Meaningful Human Control as Reason-Responsiveness: The Case of Dual-Mode Vehicles.' *Ethics and Information Technology* 22, no. 2 (2020): 103–15. https://doi.org/10.1007/s10676-019-09519-w.

Mecacci, Giulio, Daniele Amoroso, Luciano Cavalcante Siebert, David Abbink, Jeroen van den Hoven, and Filippo Santoni de Sio. *Research Handbook on Meaningful Human Control of Artificial Intelligence Systems*. Cheltenham: Edward Elgar Publishing, forthcoming.

National Transportation Safety Board. 'Collision between Vehicle Controlled by Developmental Automated Driving System and Pedestrian, Tempe, Arizona March 18, 2018.' Highway Accident Report NTSB/HAR-19/03. Washington, DC, 2019.

Nozick, Robert. *Philosophical Explanations*. Cambridge, MA: Harvard University Press, 1981.

Parkin, Simon. 'Killer Robots: The Soldiers That Never Sleep.' 16 July 2015. https://www.bbc.com/future/article/20150715-killer-robots-the-soldiers-that-never-sleep.

Pereboom, Derk. *Living Without Free Will*. Cambridge: Cambridge University Press, 2001. https://doi.org/10.1017/CBO9780511498824.

Ribeiro, Marco Tulio, Sameer Singh, and Carlos Guestrin. '"Why Should I Trust You?" Explaining the Predictions of Any Classifier.' In *Proceedings of the 22nd ACM SIGKDD International Conference on Knowledge Discovery and Data Mining - KDD '16*, 1135–44. New York: ACM Press, 2016. https://doi.org/10.1145/2939672.2939778.

Roff, Heather M., and Richard Moyes. 'Meaningful Human Control, Artificial Intelligence and Autonomous Weapons.' 2016.

Roorda, Mark. 'NATO's Targeting Process: Ensuring Human Control Over (and Lawful Use of) 'Autonomous' Weapons.' In *Autonomous Systems: Issues for Defence Policymakers*, edited by Andrew P. Williams and Paul D. Scharre. HQ Sact, 2016.

Santoni de Sio, Filippo, and Jeroen van den Hoven. 'Meaningful Human Control Over Autonomous Systems: A Philosophical Account.' *Frontiers in Robotics and AI* 5 (February 2018): 15. https://doi.org/10.3389/frobt.2018.00015.

Saxon, Dan. 'Autonomous Drones and Individual Criminal Responsibility.' In *Drones and Responsibility: Legal, Philosophical, and Socio-Technical Perspectives on the Use of Remotely Controlled Weapons*, edited by Ezio Di Nucci and Filippo Santoni de Sio, 17–46. London and New York: Routledge, 2016.

Sharkey, Noel. 'Automated Killers and the Computing Profession.' *Computer* 40, no. 11 (November 2007): 122–24. https://doi.org/10.1109/MC.2007.372.

Smiley, Lauren. 'The Legal Saga of Uber's Fatal Self-Driving Car Crash Is Over.' *Wired*, 28 July 2023. https://www.wired.com/story/ubers-fatal-self-driving-car-crash-saga-over-operator-avoids-prison/.

Stilgoe, Jack. 'Who Killed Elaine Herzberg?' In *Who's Driving Innovation?*, 1–6. Cham: Palgrave Macmillan, 2020. https://doi.org/10.1007/978-3-030-32320-2_1.

Stilgoe, Jack, Richard Owen, and Phil Macnaghten. 'Developing a Framework for Responsible Innovation.' *Research Policy* 42, no. 9 (2013): 1568–80. https://doi.org/10.1016/j.respol.2013.05.008.

Timpe, Kevin. 'Tracing and the Epistemic Condition on Moral Responsibility.' *The Modern Schoolman* 88, no. 1 (1 April 2011): 5–28. https://doi.org/10.5840/schoolman2011881/22.

Van den Hoven, Jeroen. 'Moral Responsibility, Public Office and Information Technology.' In *Public Administration in an Information Age: A Handbook*, Vol. 579. edited by I. Th. M. Snellen and Wim B. H. J. van de Donk Amsterdam: IOS Press, 1998.

———. 'Value Sensitive Design and Responsible Innovation.' In *Responsible Innovation: Managing the Responsible Emergence of Science and Innovation in Society*, edited by Richard Owen, John Bessant, and Maggy Heintz, 75–83. Hoboken, NJ: John Wiley & Sons, Ltd, 2013.

Van Inwagen, Peter. *An Essay on Free Will*. Oxford: Clarendon Press, 1983.

Veluwenkamp, Herman. 'Reasons for Meaningful Human Control.' *Ethics and Information Technology* 24, no. 4 (23 November 2022): 51. https://doi.org/10.1007/s10676-022-09673-8.

Vignard, Kerstin. 'The Weaponization of Increasingly Autonomous Technologies: Considering How Meaningful Human Control Might Move Discussion Forward.' *UNIDIR Resources*, no. 2 (2014).

Vincent, James. 'Have Autonomous Robots Started Killing in War? The Reality Is Messier than It Appears.' *The Verge*, 2022. https://www.theverge.com/2021/6/3/22462840/killer-robot-autonomous-drone-attack-libya-un-report-context.

Wagner, Ben. 'Liable, But Not in Control? Ensuring Meaningful Human Agency in Automated Decision-Making Systems.' *Policy & Internet* 11, no. 1 (March 2019): 104–22. https://doi.org/10.1002/poi3.198.

Yudowsky, Eliezer. 'Artificial Intelligence as a Positive and Negative Factor in Global Risk.' In *Global Catastrophic Risks*, edited by Nick Bostrom and Milan M. Ćirković, 308–345. New York: Oxford University Press, 2011.

9 The social dimensions of Meaningful Human Control over AI

9.1 Throwing like a girl

In a 1952 study on the psychological implications of the human upright posture, the psychologist Erwin Straus noticed 'the remarkable difference in the manner of throwing of the two sexes.'[1] Unable to find a satisfactory anatomical explanation for the phenomenon, and excluding that it could be an acquired difference as the difference was already observed in children as young as five, Straus concluded that there is a specific, biologically grounded, 'feminine attitude' in relation to world and to space.[2] In the 1980 article 'Throwing like a Girl,' the feminist philosopher Iris Young criticises Straus' approach. By noticing that boys use their whole body to throw while girls just use one arm, and by relying on the phenomenological approach of the French philosopher Merleau-Ponty, she instead proposed a different phycological explanation: girls have an 'inhibited intentionality,' that is, their capacity to throw is crucially affected by their self-imposed judgement that they cannot do it.[3] In this chapter, I will argue that something similar may happen with human control over technology. Some people or groups may be less able to control technology due to social and psychological factors depending on power structures and related identities and roles.

Technological processes may not sufficiently reflect the voices of people from oppressed groups. And these people may develop, like the girls of Straus' experiment, some socially induced inhibition in their capacity to use technology. A theory of Meaningful Human Control should carefully attend to these elements. This is the main claim of this chapter: Meaningful Human Control has a distinctive social dimension, one that is dependent on structures of social powers, and this dimension must be made explicit in the theory to prevent the further perpetuation of power structures through technological design.

The social dimension of control over technology is not a completely new topic. In the first part of this chapter, I present two theories that have addressed it before, though in different terms: the seminal work of David

DOI: 10.4324/9781003303244-12

Collingridge on the 'social control of technology' from 1980[4] and the already mentioned, more recent Responsible Innovation approach to technological development.[5]

In the second part of the chapter, I will briefly look back at the theory of Meaningful Human Control as reason-responsiveness presented in Chapter 8 and compare it with Collingridge and Responsible Innovation. Meaningful Human Control broadens the scope of the previous approaches by explicitly addressing the issue of human control over AI systems, in addition to the issue of societal control over the innovation process. However, in its current formulation, Meaningful Human Control theory does not sufficiently address the question of how social and power structures may influence the control over AI systems by socially underrepresented or oppressed groups.

To fill this gap, in the third and last part of the chapter I will introduce the concept of 'epistemic injustice,' coming from feminist literature on moral agency and responsibility. Epistemic injustice happens when oppressive social power structures, norms, and practices make people from oppressed groups (a) less able to participate in the formation of the concepts that shape collective moral and social life, thereby perpetuating their conditions of oppression and exclusion, and (b) less able to exercise their (moral) capacity, because of the psychological burden of the stereotypes they are victim of. On the basis of these concepts, two social pre-conditions of Meaningful Human Control are identified, corresponding to its two original conditions of tracking and tracing: (1) (tracking) systems that include artificial intelligence (henceforth AI) should track reasons whose formation is not determined by social (epistemological) processes unduly influenced by social hierarchies; (2) (tracing) people should not only possess the relevant individual psychological capacities (technical and moral) but also not be burdened with social identities that may affect their performance in the interaction with the system.

9.2 Collingridge's *Social Control of Technology*

Ask technologists to build gadgets which explode with enormous power or to get men to the moon, and success can be expected, given sufficient resources, enthusiasm and organization. But ask them to get food for the poor; to develop transport systems for the journeys which people want; to provide machines which will work efficiently without alienating the men who work them; to provide security from war, liberation from mental stress, or anything else where the technological hardware can fulfil its function only through interaction with people and their society, and success is far from guaranteed. [there is] a huge disparity between our technical competence and our understanding of how the fruits of this competence affect human society.[6]

In his 1980 book *The Social Control of Technology*, Collingridge argues that the problem of control over technology is not technical, but social and political. To achieve control over technology, he argues, we should move away from theories of control in engineering – which he considers to be too limited in their scope – to decision theory in policymaking.

Famously, Collingridge presented the problem of social control over technology in the form of a dilemma. While introducing a new technology in society, engineers and policymakers are not able to predict its real impact sufficiently in advance, when they can still govern it effectively (for example, who could imagine all the possible social consequences of the internet *when it was being first introduced*?). But when the impact is clearly recognisable, and it turns out to be different than expected or desired, it is usually too late to do anything about it (for example, think how difficult it is to regulate data acquisition on the internet *now*). The technology has already become embedded in social practices, vested interests will work for its perpetuation, etc. Imagine trying to 'recall' the internet due to its impact on privacy. This is Collingridge's famous dilemma of control.

He claims that most engineers and policymakers who have dealt with the dilemma have rushed to address the first part of the dilemma, which is the lack of sufficient predictability of the technological process. New tools for modelling and predictions have been studied and developed. Collingridge thinks this is not the way to go. He explicitly refers to the so-called 'fallibilist' approach in the philosophy of science (by Karl Popper and others) and claims that mistakes in predictions about the innovation process are simply ineliminable. For as much as we sharpen our prediction tools, we will never be able to fully predict the social consequences of technological innovations. A better way out of the dilemma, Collingridge argues, is then to focus on the second part: the difficulty of governing technologies after they are introduced. The key to do so is by changing our understanding of the decision-making process. In a highly compelling quote, he states that a

> decision is not just the selection of an option, but this plus a search for information in the future and the ability and resolve to act on this information if it is found. *The decision is, in short, a process and not a point.*[7]

To address the dilemma of control over technology, we need decisions that are corrigible and insensitive to so-called lock-in effects. Knowing in advance that we *will* make mistakes, we should not take decisions that cannot be reversed or corrected. Also, mechanisms of accountability should be built into the innovation process to ensure that those who have decisional power maintain the ability to respond to their (wrong) decisions and actions.[8]

A second, related, element in Collingridge's strategy to cope with uncertainty is the promotion of pluralism among expert opinions. Since no single expert can fully account for or predict the future development of technology and its societal impact, they will fill their ignorance by interpreting the available information. This interpretation will be inevitably guided by normative commitments, resulting in, indeed, an expert *opinion*. Therefore, no single expert opinion should be given full authority over the future of technological development, as none is infallible. Expert opinions should be constantly open to scrutiny, contestation, and revision. Also, the opinions of the expert minorities should be given sufficient space.

As we will see later while discussing the role of power structures in unduly influencing the formation of socially shared concepts, there is great merit in Collingridge's focus on pluralism and accountability as an antidote to the power structures negatively affecting the quality of decision-making processes by enabling a single, potentially partial or biased, interpretation of technological facts and futures to be imposed on others.

However, by coming from the mainstream philosophy of science and decision theory, this account still largely relies on experts as the main actors of knowledge and control. Collingridge does not seem to give much space to the possibility of a broader range of 'non-experts' potentially shaping decisions over technological processes and technology itself.

In fact, Collingridge is quite dismissive about the relevance of the plurality of values among users of technology and the ways their values may change or develop with time in the interaction with technology. He seems to suggest that experts should not waste time on reflecting or considering users' different values and attitudes towards technology, as these tend in the end to be uniform and stable across time. Experts should rather focus on analysing and making predictions about technological development. In his words:

> The sort of values which figure in [the policy] debate are generally so mundane and ordinary that only the wildest eccentric would want to question them. These values are shared by the parties to the debate, but can be expected to be also held by any of the experts they consult.' [Therefore,] [t]he expert's task is to offer and defend interpretations of the data, [and] [t]he interpretation is the key item, it is this which is disputed, not the value judgements.[9]

Collingridge identified an important social dimension of control over technology and his insights into accountability and pluralism are still considered valuable.[10] At the same time, social science and philosophical research on the societal control of technological innovation has advanced, looking beyond the role of experts and proposing a more complex and dynamic relationship between societal actors and technological innovation.[11]

9.3 Democratising the governance of technology: Responsible Innovation

More recently, philosophers of technology have tried to develop the idea of social control over technology in a way that gives not only experts but a broader range of stakeholders more control over technological processes. Ibo van de Poel follows Collingridge and his philosophical inspiration in Popper's fallibilism in proposing decision-making processes that are somehow controllable. To achieve this, taking direct inspiration from John Dewey's pragmatist approach, van de Poel proposes to consider the introduction of technology in society as a 'social experiment.'[12] Whenever a new technology is introduced, van de Poel argues, citizens and people are *de facto* subjects of a social and technological experiment, which bears opportunities and risks, and whose results are never fully predictable. Like scientific experiments, socio-technological experiments should also comply with some standards of ethical research. A similar idea had been presented by Martin and Schinzinger,[13] according to whom a social experiment is ethically acceptable only if citizens-subjects give their informed consent. Van de Poel thinks that asking for consent is at the same time too little and too much to make technological experiments ethically acceptable. Too little, as there is more than consent to the acceptability of an experiment – people may have many reasons to consent to things that are, all things considered, not in their interests. Too much as it is difficult to imagine ways to have all citizens express their free and informed consent on a technological innovation.

Van de Poel therefore proposes a framework that connects social control over technology with some well-established principles for ethical research, that is those of medical ethics, embedded in documents such as the Nuremberg Code (NC), the Helsinki Declaration (HD), and the Common Rule (CR). The result is a list of 16 principles that should guide socio-technological experimentation to ensure that it complies with the four basic principles of medical ethics: non-maleficence, beneficence, autonomy of subjects, and distributive justice. As interesting and original as this proposal may be, it still presents experts and policymakers as the *agents* of technology who can shape and steer technological progress and socio-technical systems, and all other societal actors as *subjects*, whose rights and interests should be protected, and who can control technology only in the sense of monitoring, deliberating, criticising, and possibly stopping.

Another approach that has tried to give common people a greater say in technological processes is the so-called Responsible Innovation movement, which started in Europe in the 2010s. Different aspects of value alignment are captured in the Responsible Innovation literature under the idea of 'responsiveness.' To ensure that a technological process is societally beneficial, at least three forms of responsiveness should be pursued: (a) between innovators and stakeholders,[14] (b) between the innovation process and changing information environment and values,[15] and (c) among stakeholders.[16] Different authors within the Responsible Innovation tradition tend to emphasise different aspects.

Within the confining field of the so-called Technology Assessment, Grunwald stresses the importance of the active inclusion of diverse interpretations of the meaning of (emerging) technologies for the future of society. He relies on Dewey, Habermas, and hermeneutics to support the importance of this inclusive deliberation.[17]

One important idea behind the Responsible Innovation and the Technology Assessment approaches is that the democratisation of the innovation process is not only ethically desirable but also scientifically beneficial. Including and engaging stakeholders' views and concerns is not only required as a matter of ethics or justice, since they will be affected by the introduction of technology, like in van de Poel's idea of technology as a social experiment. The inclusion of stakeholders should also be pursued for epistemological reasons, as it can crucially contribute to a more 'robust' knowledge.[18] As already pointed out by Collingridge, the natural sciences and engineering do not possess all the knowledge required to understand and govern the introduction of technology in society. Responsible Innovation proponents argue that social science – narrowly construed – doesn't either. Some have argued that we have entered the age of 'post-normal science' in which complex scientific and societal challenges – fighting cancer, coping with pandemics, and governing AI – cannot be effectively addressed with traditional scientific tools only, but require mobilising and connecting different sources of knowledge that are distributed across different stakeholders and groups: scientists; policymakers; and also citizens, activists, professionals.[19]

Responsible Innovation approaches have greatly contributed to stressing the importance of the distribution of social power in innovation processes to make technology more responsive to the values and needs of people. In this sense, they are an important model for Meaningful Human Control to look at.

At the same time, they have also been criticised for holding an overly idealised view of the way this deliberative process among stakeholders may take place. They have not sufficiently considered the 'politics in deliberation,' in other words, the concrete social processes and power structures involved in deliberation. 'How do actors "co-create" outcomes? How do they deliberate? On whose terms is participation (i.e. deliberation) established, and why?'[20]

As I will argue, this is a weakness shared by the current formulations of the theory of Meaningful Human Control as reason-responsiveness, to which I now turn.

9.4 Meaningful Human Control over AI: current formulations and their limitations

Recent developments in AI have prompted a renewed interest in the concept of control over technology. Distinguished scientists such as Stephen Hawking and, in the field of AI, Stuart Russell have identified achieving control over AI as one of the biggest and most urgent challenges of the next decades.[21] Hawking, Russell, and others concerned about the risks of 'singularity'[22] or 'superintelligence'[23] are mostly looking at a long-term perspective in which AI may move from its current capacity to perform specific tasks to that of developing more human-like forms of intelligence but in a more powerful way, thereby potentially threatening the existence of humankind itself. Looking at the shorter run, heated political debates have been prompted on the possibility of (losing) human control of specific AI applications. In the previous chapter, for instance, I have discussed the ethical and political concerns about that AI-driven military technology – so-called 'killer robots' or 'autonomous weapon systems' – may slip out of human control. It is within these debates that the concept of 'Meaningful Human Control' over AI first emerged.[24] The concern is that not the innovation process but the technical systems themselves may elude human control, due to their intelligent, learning features and the specific complexity and unpredictability of their interactions with users.

In the previous chapter I have proposed a philosophical account of Meaningful Human Control, taking reason-responsiveness as a general framework for designing AI systems that remain responsive to the relevant reasons of – depending on the context – the designers, users, regulators, or other human agents interacting with the AI systems. The framework aims to guarantee the alignment between AI-based systems and different human agents' reasons – values, principles, and intentions – as well as to avoid so-called responsibility gaps, situations in which someone is wronged or unjustly harmed but it is difficult to identify with clarity who should be held responsible, in various senses, for that wrong or harm.[25]

The theory relies on a comparison with the attribution of moral responsibility for everyday human actions, as defined by Fischer and Ravizza.[26] Inspired by the Value-Sensitive Design approach,[27] the Meaningful Human Control approach reinterprets Fischer's and Ravizza's conditions for moral responsibility as conditions to keep responsibility over AI systems and translates these two conditions into two corresponding design requirements to keep AI systems under Meaningful Human Control. The two requirements are called 'tracking' and 'tracing.' They can be summarised as follows:

> Tracking requires that the socio-technical system – i.e. the whole combination of technical, human, and organisational elements – is designed to demonstrably respond to the relevant reasons of the relevant agents and to the relevant facts in the environment. Tracing requires that the socio-technical system is designed so that for each of the (relevant) actions of the system, it is possible to identify at least one human agent (ideally more) along the chain of design, development, and use that possesses both (i) sufficient knowledge of the capabilities and limitations of the system and (ii) sufficient moral awareness of, and capacity to comply with, her role as potential target of legitimate response for the behaviour of the system. Tracking requires the alignment of the system with the values, reasons, and intentions of the relevant human agents; tracing requires the alignment of the system with the capacities of the relevant human agents.[28]

The Meaningful Human Control framework expands Collingridge's and the Responsible Innovation's approach to societal control over technology in two ways.

First, by being grounded in the theory of (individual) moral responsibility, Meaningful Human Control theory explicitly invites recognising the different actors interacting with technology – designers, regulators, users, and others – not only as *stakeholders* whose interests should be reflected *in the innovation process* but also as potential (individual) *responsible agents* who, given certain conditions, are in the position to *steer an individual socio-technical system* in a desirable direction. So, it shifts attention from control over the process to control over the systems.

Second, by being inspired by Value-Sensitive Design, its conceptual and normative requirements – tracking and tracing – are cast in terms that could be used to design an AI system for controllability and responsibility, that is, to empower specific agents in the system and keeping them responsible for its behaviour.[29]

However, this strong focus on Meaningful Human Control and responsible agency by actors within a given socio-technical system – as opposed to a broader focus on the technological process – brings the risk of ignoring some important lessons coming from Collingridge and the Responsible Innovation. There are three related risks:

First, by mainly focusing on control over the behaviour of specific technological (AI) systems – as opposed to the control over the technological process – Meaningful Human Control may risk – against the spirit of

Responsible Innovation – to close the space to the participation of some agents and the inclusion of some views, thereby perpetuating social injustice and exclusion. Once control is framed and defined as control over a specific socio-technical system such as, for instance, self-driving cars, the reasons of those who do not want these systems at all, or who think technological investments should go elsewhere, may not be easily included in the control equation. Too much attention to control – in a purely engineering sense – may indeed lead to less inclusivity of agents and a less responsible ('responsive'), democratic innovation process.

Second, the preliminary exclusion of some voices from the list of the relevant agents may give controllers of technology the illusion that their technology and technological process are more predictable or linear than they really are, thereby running the risk – against the advice of Collingridge – of not including sufficient measures to govern the intrinsic uncertainty of the technological process and the changing nature of values. Or, in the words of Sheila Jasanoff, feeding 'technologies of hubris' rather than 'technologies of humility.'[30] The ideal of designing socio-technical systems for control and responsibility may create an excessive illusion of control in designers and society, one that obfuscates the intrinsic openness, uncertainty, and unpredictability of the technological development. Instead, the inclusion of critical voices raising radical concerns and broader objections about the technology may keep developers and society more alert to the possible unwanted or even unpredictable directions that technology may take.

Finally, like the concept of moral responsibility by which it is inspired – Fischer and Ravizza's reason-responsiveness – the Meaningful Human Control theory does not explicitly address the issue of social and power structures as potential obstacles to responsible agency and control. More specifically, feminist literature has highlighted the epistemological dimension of social oppression. Persons who happen to be part of oppressed groups are often also less able to contribute to forming the shared concepts or ideas that will shape moral and social practices. This 'epistemic injustice,' as it is called, can negatively affect, among other things, their capacity for social agency, that is, their power to shape the environment in which they live. It can also directly affect people's capacity to deploy their capacities, by creating psychological inhibitions through the burden of stereotypes. This social dimension of moral agency is not sufficiently reflected in many theories of moral responsibility, including the one grounding our Meaningful Human Control approach.

So, while proposing to enrich the Responsible Innovation approach by explicitly giving human agents more power to steer AI systems in desirable directions, the Meaningful Human Control approach has also so far failed to include in its theory measures to address the negative effects of concrete social dynamics and power structures. These may limit the opportunity and capacities of some actors to exert their control.

To overcome these limitations, in the next section, I will propose to integrate the Meaningful Human Control approach over AI systems with two pre-conditions, based on insights on 'epistemic injustice' from feminist social epistemology and moral philosophy. This is one way to make Meaningful Human Control more inclusive and sensitive to social justice.

9.5 The social dimensions of Meaningful Human Control

9.5.1 The social dimensions of moral responsibility

The theory of Meaningful Human Control as reason-responsiveness is grounded in one specific theory of moral responsibility: Fischer and Ravizza's theory of responsibility as reason-responsiveness.[31] So, *if* Fischer and Ravizza are right about the conditions for people to remain in moral control of and responsible for their actions, i.e. actions considered in isolation from the mediation of complex technological systems, *then* we must just adjust and reframe these conditions to define the requirements for human agents to remain in control of and responsible for the actions mediated by technological systems.

However, recent literature on moral responsibility has convincingly argued that Fischer and Ravizza's theory of moral responsibility is in fact incomplete. And so arguably is the theory of Meaningful Human Control as reason-responsiveness.[32] This critique to Fischer and Ravizza comes from theorists of moral responsibility inspired by studies in moral psychology, social epistemology, and (feminist) moral philosophy.

These studies have emphasised the following two general points:

Reason formation is sensitive to power structures. Fischer and Ravizza hold that control and responsibility depend on the capacity to respond to the relevant (moral) reasons but do not discuss the social mechanisms through which (moral) reasons acquire relevance in society – as it were, who decides which reasons are relevant. This is an important gap, because, as a matter of fact, (moral) reasons acquire relevance – i.e. visibility and force – in society, through processes that are heavily shaped by positions and dynamics of power. Therefore, societies with strong imbalances of power will not only promote injustice in the distribution of goods and offices between dominant and oppressed groups, for instance, unjustified inequality in salaries or access to economic or political positions. They will also create forms of 'epistemic injustice', that is unjustified inequalities in the *possibility to contribute to the formation of shared knowledge*.[33] Epistemic injustice may take various forms. Oppressed groups may not enjoy the same level of credibility in society as other groups, like in Harper Lee's *To Kill a Mockingbird*, when Tom Robinson, the young Black man accused of rape of the young white woman Mayella Ewell, cannot convince the (white) jurors of his innocence because he feels his version of the facts – that he was with Mayella to help her with her chores and that he ran away after she tried to kiss him – would be met with incredulity or disdain. On the other hand, Mayella's version of the facts – that Tom raped her and then ran away because of his guilt consciousness – better matches the expectations of the jurors and so is readily accepted by them.[34] Here, the acceptance of certain facts as true by a community crucially depends on the position of power of those who present them. This is called 'testimonial injustice.' Other times, people from oppressed groups do not have the power to have their *interpretations* of facts recognised in society, like when the behaviour of men (employers) that are perceived as harassing by women (employees) is dismissed as just light, innocent jokes by those men, and it is very difficult for women to make their interpretation of facts being accepted in society.[35] This is called 'hermeneutic injustice.' Finally, members of disadvantaged or oppressed groups may have a hard time delivering the desired meaning and effects to their speech acts, like when a woman floor manager is not able to get things done by her collaborators, because her orders to carry out certain tasks are systematically interpreted as requests – as opposed to orders – and treated as such. This is called 'discursive injustice.'[36] In general, due to epistemic injustice, the relevance of reasons in a society is determined through social and epistemological processes where powerful groups are overrepresented and less powerful or oppressed groups are under- or not represented. What looks like *relevant* reasons in society are in fact often *dominant* reasons.

Human capacities are affected by social identities. Also, Fischer and Ravizza's and other recent theories of moral responsibility as reason-responsiveness do not address the issue of how people's capacities to understand and endorse the mechanisms leading to their action are influenced by social dynamics and structures of power.[37] Social identities – the way people are perceived by others and themselves in a community due to their gender, racial, economic, and other profiles – may affect people's capacities. As Miranda Fricker explains, the subject of a stereotype 'may actually resemble the prejudicial stereotype working against her.'[38] A woman may interiorise the idea of her intellectual inferiority promoted in a patriarchal society and stop developing her intellectual skills, thereby actually losing them. Also, empirical studies show that

> when individuals in a situation (e.g., a job interview) or doing a task (e.g., an IQ or math test) are exposed to negative stereotypes about the typical performance of members of their social groups (e.g., women, African Americans) in situations or tasks of that kind, they perform significantly less well.[39]

9.5.2 The social pre-conditions of Meaningful Human Control

The first condition for Meaningful Human Control ('tracking') requires AI systems to be responsive to the relevant reasons of the relevant agents. The theory of Meaningful Human Control as reason-responsiveness holds that what counts as a relevant reason and a relevant agent in a system cannot be defined by the theory itself.[40] These are highly context-dependent assessments that cannot be included in a generic framework. For instance, the relevant reasons and agents in a transport system that includes automated vehicles are different than those of a digital commercial platform run by AI or a healthcare system that includes AI-based diagnostic tools. Also, there may be legitimate disagreement on what should count as a relevant reason and agent in each of these domains. New reasons and agents may emerge over time, which cannot be included in advance in the theory of Meaningful Human Control. These must be discussed and negotiated.

However, in the light of the above discussion on the social dimensions of moral responsibility, the theory of Meaningful Human Control should arguably include some extra requirements. Especially when design-oriented, a theory of Meaningful Human Control should be cognisant of the structures of power that could make the social negotiation about relevant reasons unfair and prevent people from being able to develop or exercise the relevant capacities. If these factors are not considered, socio-technical systems risk only tracking the *dominant* reasons of the dominant agents and not all the *relevant* ones; Meaningful Human Control should include a more explicit awareness of the moral psychology of reason exchange and the extent to which this exchange is affected by social hierarchies and power structures. By not addressing these social factors, Meaningful Human Control may de facto deprive some agents and their reasons from being recognised in the system design. Instead, by actively contrasting these structures of power, new relevant agents and reasons may emerge and be included in the system's design and operation.

Therefore, I propose that the tracking condition for Meaningful Human Control explicitly includes the following social pre-condition:

> *First social pre-condition for Meaningful Human Control* ('tracking'): socio-technical systems that include AI should track reasons whose formation is not heavily influenced by social (epistemological) processes unduly influenced by unjust social hierarchies. There should not be major social or political structures preventing groups from contributing to the formation of the shared reasons in the social communities where the technology will be developed, introduced, and utilised.

Since unjust social hierarchies cannot arguably be eliminated, this pre-condition must be read as a criterion, not a binary requirement: the lower the presence of (epistemic) injustices, the more meaningful the human control over technology.

Consider, as an example, algorithmic bias and related injustice. A growing body of academic and policy literature has already emphasised the risks of so-called algorithmic bias for social justice. The algoriths behind Google's search engines have been shown to reinforce demeaning stereotypes of darker-skinned young women.[41] Facial recognition techniques by Amazon and others have shown similar patterns.[42] The AI systems reflect some biases present in society and in existing data and reinforce these biases by incorporating them in systems that heavily contribute to shaping people's mental categories, practices, and behaviours. One obvious way of addressing this is merely technical: a better design of algorithms and, in particular, careful consideration of the inclusivity and representativity of (training) data sets used by the machines. Another often-mentioned remedy is increasing the social inclusivity of the design process, for example, by having more representatives of underrepresented societal groups in the design, governance, or regulation groups, or through a better individual awareness of the impact of implicit bias.[43]

The social pre-condition for 'tracking' presented above suggests yet another level of intervention. People included in the design process may still not have the power to make their voice heard, or their voice may not be considered in the same way in which other voices are. Therefore, diversity and inclusivity may not be enough. Specific mechanisms and procedures are needed to detect and counterbalance languages, practices, and patterns of behaviours reflecting unjust social epistemic processes due to social structures of power within these groups. Also, research and design groups may proactively adopt specific design methods to neutralise the impact of existing social hierarchies in the design of technology. What these methods would look like heavily depends on the specific community and technology.

A similar point can be made about the second condition for Meaningful Human Control: 'tracing.' According to 'tracing' there must always be within a socio-technical system at least one and ideally more human agents along the chain of design, development, and use, who possess both (i) sufficient knowledge of the capabilities and limitations of the system and (ii) sufficient moral awareness of, and capacity to comply with, their role as a potential target of legitimate response for the behaviour of the system.

Meaningful Human Control theory explicitly recognises the motivational and situational element of human capacity and explicitly invites studying and considering some psychological factors that can affect the capacity to fulfil one's role in the interaction with technology, e.g. motivational limitations in supervision tasks and others.[44] On the positive side,

Meaningful Human Control also emphasises the potential capacitating role of practices of responsibility. In Chapters 7 and 8, I have insisted on the importance of enhancing various agents' capacities and motivation to fulfil their responsibility in relation to the development and use of AI (thereby avoiding what I called 'active responsibility gaps'). This can be done, among others, through new forms of education and training (including the so-called digital literacy), the creation of spaces to promote a better awareness on the agents' (professional) role, and the promotion of activities and practices to foster the creation of new professional identities.[45] In this chapter, I propose to also explicitly consider the influence of social identities shaped by unjust power structures in the formation of capacities to interact with technology.

Therefore, the tracing condition for Meaningful Human Control should explicitly include also the following social pre-condition:

> *Second social pre-condition for Meaningful Human Control* ('tracing'): to be able to take moral responsibility for the behaviour of socio-technical systems, people should not only possess the relevant individual psychological capacities (technical and moral), as measured in isolation from social context, but also not be burdened with social identities that may affect their performance as moral agents in the interaction with the systems.

Consider, as an example, the interaction with self-driving cars. It has been argued that automation may create a gap between the normative expectations imposed on certain actors and their actual capacity to fulfil them, thereby creating the risk of scapegoating people for mistakes they lacked the capacity to prevent and that are rather to be attributed to a bad design of the human–machine interaction. In a classic example, a driver of a partially automated vehicle is imposed an obligation to supervise the system and to prevent accidents. As the task they are assigned is beyond their (psychological) capacities, they find themselves in a so-called 'moral crumple zone,'[46] a place where blame and liability will unfairly be attributed to them simply because of their position in the system. To prevent this problem, many – including the supporters of Meaningful Human Control as reason-responsiveness – have proposed one or more of the following: that the design of the human–machine interaction closely tracks the existing psychological capacities of (individual) users as reflected in psychological studies;[47] that people's capacities are trained to comply with the new obligations before these obligations are imposed;[48] and that users are made more aware of their (moral) responsibility in the interaction with technology.[49]

The social pre-condition for tracing proposed above suggests another level of analysis and intervention. In 'Throwing Like a Girl,' Young claimed that girls participating in Erwin Straus' experiments had an 'inhibited intentionality,' that is, their capacity to throw was crucially affected by their self-imposed judgement that they could not do it.[50] Recent psychological studies show that people may (temporarily) lose their capacities under the pressure of prejudice and stereotypes. For instance, people from disadvantaged groups – other capabilities being equal – may be less able than others to challenge the authority of a manager due to an interiorised sense of inferiority. With the introduction of AI and robot systems in the workplaces, these same people may also be less able to challenge the decision or the behaviour of an AI system.

The social pre-condition for tracing requires that these social obstacles to acquiring, maintaining, or exercising (moral) capacities are also considered in the design and deployment of AI technologies. In other words, it requires that social identities – gender, race, economic status, age, etc. as opposed to other general psychological skills – do not affect people's capacities to responsibly interact with AI systems. And that technical systems do not perpetuate and reinforce social exclusion, by giving meaningful control only to the socially powerful humans.

Notes

1 Erwin W. Straus, 'The Upright Posture,' *The Psychiatric Quarterly* 26, no. 1 (1 January 1952): 529–61, https://doi.org/10.1007/BF01568490.
2 Straus, 553.
3 Iris Marion Young, 'Throwing Like a Girl: A Phenomenology of Feminine Body Comportment Motility and Spatiality,' *Human Studies* 3 (1980): 146–47.
4 David Collingridge, *The Social Control of Technology* (London: Frances Printers, 1980).
5 Jack Stilgoe, Richard Owen, and Phil Macnaghten, 'Developing a Framework for Responsible Innovation', *Research Policy* 42, no. 9 (2013): 1568–80, https://doi.org/10.1016/j.respol.2013.05.008.
6 Collingridge, *The Social Control of Technology*, 15.
7 Collingridge, 30–31 (emphasis added).
8 Audley Genus and Andy Stirling, 'Collingridge and the Dilemma of Control: Towards Responsible and Accountable Innovation,' *Research Policy* 47 (2018): 61–69, https://doi.org/10.1016/j.respol.2017.09.012.
9 Collingridge, *The Social Control of Technology*, 192.
10 Genus and Stirling, 'Collingridge and the Dilemma of Control: Towards Responsible and Accountable Innovation.'
11 Arie Rip, 'Processes of Technological Innovation in Context – and Their Modulation,' in *Relational Practices, Participative Organizing*, eds. Chris Steyaert and Bart van Looy (Bingley: Emerald, 2010), 199–217; Olya Kudina and Peter Paul Verbeek, 'Ethics from Within: Google Glass, the Collingridge Dilemma, and the Mediated Value of Privacy,' *Science, Technology, & Human Values* 44, no. 2 (21 August 2018): 291–314, https://doi.org/10.1177/0162243918793711.

12 Ibo van de Poel, 'An Ethical Framework for Evaluating Experimental Technology,' *Science and Engineering Ethics* 22, no. 3 (1 June 2016): 667–86, https://doi.org/10.1007/S11948-015-9724-3/TABLES/3.
13 Mike W. Martin and Roland Schinzinger, *Ethics in Engineering*, 4th ed. (Boston: McGraw-Hill, 2005).
14 René von Schomberg, 'Towards Responsible Research and Innovation in the Information and Communication Technologies and Security Technologies Fields' (Luxemburg: Publications Office of the European Union, 2011), https://doi.org/10.2777/58723; Stilgoe, Owen, and Macnaghten, 'Developing a Framework for Responsible Innovation.'
15 Stilgoe, Owen, and Macnaghten, 'Developing a Framework for Responsible Innovation'; Stevienna De Saille, 'Innovating Innovation Policy: The Emergence of "Responsible Research and Innovation",' *Journal of Responsible Innovation* 2, no. 2 (4 May 2015): 152–68, https://doi.org/10.1080/23299460.2015.1045280; Jeroen van den Hoven, 'Value Sensitive Design and Responsible Innovation,' in *Responsible Innovation*, eds. Richard Owen, John Bessant, and Maggy Heintz (Chichester: John Wiley & Sons, Ltd, 2013), 75–83, https://doi.org/10.1002/9781118551424.ch4.
16 Von Schomberg, 'Towards Responsible Research and Innovation in the Information and Communication Technologies and Security Technologies Fields.'
17 Armin Grunwald, *Technology Assessment in Practice and Theory* (London and New York: Routledge, 2019).
18 Michael Gibbons, ed., *The New Production of Knowledge: The Dynamics of Science and Research in Contemporary Societies*, Reprinted (Los Angeles: Sage Publ., 2011).
19 S. O. Funtowicz and J. R. Ravetz, 'Post-Normal Science: A New Science for New Times,' *Scientific European* (October 1990): 20–22.
20 Michiel Van Oudheusden, 'Where Are the Politics in Responsible Innovation? European Governance, Technology Assessments, and Beyond,' *Journal of Responsible Innovation* 1, no. 1 (2 January 2014): 67–86, https://doi.org/10.1080/23299460.2014.882097.
21 Stuart Russell, *Human Compatible: Artificial Intelligence and the Problem of Control* (New York: Penguin UK, 2019).
22 Ray Kurzweil, *The Singularity Is near: When Humans Transcend Biology* (New York, NY: Penguin Books, 2006).
23 Nick Bostrom, *Superintelligence: Paths, Dangers, Strategies* (Oxford: Oxford University Press, 2014).
24 Article 36, 'Autonomous Weapons, Meaningful Human Control and the CCW,' (Article 36, 2014), http://www.article36.org/weapons-review/autonomous-weapons-meaningful-human-control-and-the-ccw/.
25 See Chapter 7 above and Filippo Santoni de Sio and Giulio Mecacci, 'Four Responsibility Gaps with Artificial Intelligence: Why They Matter and How to Address Them,' *Philosophy & Technology* 34, no. 4 (14 December 2021): 1057–84, https://doi.org/10.1007/s13347-021-00450-x.
26 J. M. Fischer and M. Ravizza, *Responsibility and Control: A Theory of Moral Responsibility* (Cambridge: Cambridge University Press, 1998).
27 Van den Hoven, 'Value Sensitive Design and Responsible Innovation.'
28 Santoni de Sio and Mecacci, 'Four Responsibility Gaps with Artificial Intelligence: Why They Matter and How to Address Them,' 1077.
29 For a development of this idea from an engineering perspective, see Simeon C. Calvert et al., 'A Human Centric Framework for the Analysis of Automated Driving Systems Based on Meaningful Human Control,' *Theoretical Issues in*

Ergonomics Science, 10 December 2019, 1–29, https://doi.org/10.1080/1463922X.2019.1697390; Luciano Cavalcante Siebert et al., 'Meaningful Human Control: Actionable Properties for AI System Development,' *AI and Ethics* 3, no. 1 (1 February 2023): 241–55, https://doi.org/10.1007/s43681-022-00167-3.

30 Sheila Jasanoff, 'Technologies of Humility: Citizen Participation in Governing Science,' *Minerva* 41, no. 3 (1 September 2003): 223–44, https://doi.org/10.1023/A:1025557512320.

31 Fischer and Ravizza, *Responsibility and Control*.

32 This idea was originally suggested to me by Marianna Capasso, 'Positions of Power. Freedom and Control in the Age of Artificial Intelligence' (Pisa, School of Advanced Studies Sant'Anna, 2022), https://dta.santannapisa.it/t/etd-12092021-161754/.

33 Miranda Fricker, *Epistemic Injustice: Power and the Ethics of Knowing* (Oxford: Oxford University Press, 2007).

34 Catriona Mackenzie, 'Moral Responsibility and the Social Dynamics of Power and Oppression,' in *Social Dimensions of Moral Responsibility*, eds. Katrina Hutchison, Catriona Mackenzie, and Marina Oshana (New York: Oxford University Press, 2018), 72.

35 Mackenzie, 18–19.

36 Mackenzie, 71; Rebecca Kukla, 'Performative Force, Convention, and Discursive Injustice,' *Hypatia* 29, no. 2 (2014): 441, https://doi.org/10.1111/j.1527-2001.2012.01316.x.

37 Mackenzie, 'Moral Responsibility and the Social Dynamics of Power and Oppression.'

38 Fricker, *Epistemic Injustice*, 55.

39 Mackenzie, 'Moral Responsibility and the Social Dynamics of Power and Oppression,' 65; C. M. Steele, 'A Threat in the Air. How Stereotypes Shape Intellectual Identity and Performance,' *The American Psychologist* 52, no. 6 (June 1997): 613–29, https://doi.org/10.1037//0003-066x.52.6.613.

40 Giulio Mecacci and Filippo Santoni de Sio, 'Meaningful Human Control as Reason-Responsiveness: The Case of Dual-Mode Vehicles,' *Ethics and Information Technology* 22, no. 2 (19 June 2020): 113, https://doi.org/10.1007/s10676-019-09519-w.

41 Safiya Umoja Noble, *Algorithms of Oppression: How Search Engines Reinforce Racism* (New York: New York University Press, 2018).

42 James Vincent, 'Gender and Racial Bias Found in Amazon's Facial Recognition Technology,' *The Verge*, 25 January 2019, https://www.theverge.com/2019/1/25/18197137/amazon-rekognition-facial-recognition-bias-race-gender.

43 Michael Li, 'To Build Less-Biased AI, Hire a More-Diverse Team,' *Harvard Business Review*, 26 October 2020, https://hbr.org/2020/10/to-build-less-biased-ai-hire-a-more-diverse-team.

44 Filippo Santoni de Sio et al., 'Realising Meaningful Human Control Over Automated Driving Systems: A Multidisciplinary Approach,' *Minds and Machines*, 28 July 2022, https://doi.org/10.1007/s11023-022-09608-8.

45 Jean-François Bonnefon et al., *Horizon 2020 Commission Expert Group to Advise on Specific Ethical Issues Raised by Driverless Mobility (E03659). Ethics of Connected and Automated Vehicles: Recommendations on Road Safety, Privacy, Fairness, Explainability and Responsibility* (Luxembourg: European Commission, 2020), 55–57.

46 Madeleine Clare Elish, 'Moral Crumple Zones: Cautionary Tales in Human-Robot Interaction,' *Engaging Science, Technology, and Society* 5 (23 March 2019): 40, https://doi.org/10.17351/ests2019.260.

47 Daniël D. Heikoop et al., 'Human Behaviour with Automated Driving Systems: A Quantitative Framework for Meaningful Human Control,' *Theoretical Issues in Ergonomics Science* 20, no. 6 (2 November 2019): 711–30, https://doi.org/10.1080/1463922X.2019.1574931.
48 Santoni de Sio et al., 'Realising Meaningful Human Control Over Automated Driving Systems.'
49 Bonnefon et al., *Horizon 2020 Commission Expert Group to Advise on Specific Ethical Issues Raised by Driverless Mobility (E03659). Ethics of Connected and Automated Vehicles: Recommendations on Road Safety, Privacy, Fairness, Explainability and Responsibility*, 55–56.
50 Young, 'Throwing Like a Girl: A Phenomenology of Feminine Body Comportment Motility and Spatiality,' 146–47.

References

Article 36. 'Autonomous Weapons, Meaningful Human Control and the CCW.' Article 36, 2014. http://www.article36.org/weapons-review/autonomous-weapons-meaningful-human-control-and-the-ccw/.

Bonnefon, Jean-François, David Černy, John Danaher, Nathalie Devillier, Veronika Johansson, Tatiana Kovacikova, Marieke Martens, Miloš Mladenovic, Paula Palade, Nick Reed, Filippo Santoni de Sio, Stavroula Tsinorema, Sandra Wachter, and Karolina Zawieska. *Horizon 2020 Commission Expert Group to Advise on Specific Ethical Issues Raised by Driverless Mobility (E03659). Ethics of Connected and Automated Vehicles: Recommendations on Road Safety, Privacy, Fairness, Explainability and Responsibility*. Luxembourg: European Commission, 2020.

Bostrom, Nick. *Superintelligence: Paths, Dangers, Strategies*. Oxford: Oxford University Press, 2014.

Calvert, Simeon C., Daniël D. Heikoop, Giulio Mecacci, and Bart van Arem. 'A Human Centric Framework for the Analysis of Automated Driving Systems Based on Meaningful Human Control.' *Theoretical Issues in Ergonomics Science* (10 December 2019): 1–29. https://doi.org/10.1080/1463922X.2019.1697390.

Capasso, Marianna. 'Positions of Power. Freedom and Control in the Age of Artificial Intelligence.' *School of Advanced Studies Sant'Anna* (2022). https://dta.santannapisa.it/t/etd-12092021-161754/.

Cavalcante Siebert, Luciano, Maria Luce Lupetti, Evgeni Aizenberg, Niek Beckers, Arkady Zgonnikov, Herman Veluwenkamp, David Abbink, et al. 'Meaningful Human Control: Actionable Properties for AI System Development.' *AI and Ethics* 3, no. 1 (1 February 2023): 241–55. https://doi.org/10.1007/s43681-022-00167-3.

Collingridge, David. *The Social Control of Technology*. London: Frances Printers, 1980.

De Saille, Stevienna. 'Innovating Innovation Policy: The Emergence of "Responsible Research and Innovation".' *Journal of Responsible Innovation* 2, no. 2 (4 May 2015): 152–68. https://doi.org/10.1080/23299460.2015.1045280.

Elish, Madeleine Clare. 'Moral Crumple Zones: Cautionary Tales in Human-Robot Interaction.' *Engaging Science, Technology, and Society* 5 (23 March 2019): 40. https://doi.org/10.17351/ests2019.260.

Fischer, J. M., and M. Ravizza. *Responsibility and Control: A Theory of Moral Responsibility*. Cambridge: Cambridge University Press, 1998.

Fricker, Miranda. *Epistemic Injustice: Power and the Ethics of Knowing*. Oxford: Oxford University Press, 2007.

Funtowicz, S. O., and J. R. Ravetz. 'Post-Normal Science: A New Science for New Times.' *Scientific European* (October 1990): 20–22.

Genus, Audley, and Andy Stirling. 'Collingridge and the Dilemma of Control: Towards Responsible and Accountable Innovation.' *Research Policy* 47 (2018): 61–69. https://doi.org/10.1016/j.respol.2017.09.012.

Gibbons, Michael, ed. *The New Production of Knowledge: The Dynamics of Science and Research in Contemporary Societies*. Los Angeles: Sage Publications, 2011.

Grunwald, Armin. *Technology Assessment in Practice and Theory*. London and New York: Routledge, 2019.

Heikoop, Daniël D., Marjan Hagenzieker, Giulio Mecacci, Simeon Calvert, Filippo Santoni De Sio, and Bart Van Arem. 'Human Behaviour with Automated Driving Systems: A Quantitative Framework for Meaningful Human Control.' *Theoretical Issues in Ergonomics Science* 20, no. 6 (2 November 2019): 711–30. https://doi.org/10.1080/1463922X.2019.1574931.

Jasanoff, Sheila. 'Technologies of Humility: Citizen Participation in Governing Science.' *Minerva* 41, no. 3 (1 September 2003): 223–44. https://doi.org/10.1023/A:1025557512320.

Kudina, Olya, and Peter Paul Verbeek. 'Ethics from Within: Google Glass, the Collingridge Dilemma, and the Mediated Value of Privacy.' *Science, Technology, & Human Values* 44, no. 2 (21 August 2018): 291–314. https://doi.org/10.1177/0162243918793711.

Kukla, Rebecca. 'Performative Force, Convention, and Discursive Injustice.' *Hypatia* 29, no. 2 (2014): 440–57. https://doi.org/10.1111/j.1527-2001.2012.01316.x.

Kurzweil, Ray. *The Singularity Is Near: When Humans Transcend Biology*. New York: Penguin Books, 2006.

Li, Michael. 'To Build Less-Biased AI, Hire a More-Diverse Team.' *Harvard Business Review*, 26 October 2020. https://hbr.org/2020/10/to-build-less-biased-ai-hire-a-more-diverse-team.

Mackenzie, Catriona. 'Moral Responsibility and the Social Dynamics of Power and Oppression.' In *Social Dimensions of Moral Responsibility*, edited by Katrina Hutchison, Catriona Mackenzie, and Marina Oshana, 59–80. New York: Oxford University Press, 2018.

Martin, Mike W., and Roland Schinzinger. *Ethics in Engineering*, 4th ed. Boston: McGraw-Hill, 2005.

Mecacci, Giulio, and Filippo Santoni de Sio. 'Meaningful Human Control as Reason-Responsiveness: The Case of Dual-Mode Vehicles.' *Ethics and Information Technology* 22, no. 2 (19 June 2020): 103–15. https://doi.org/10.1007/s10676-019-09519-w.

Noble, Safiya Umoja. *Algorithms of Oppression: How Search Engines Reinforce Racism*. New York: New York University Press, 2018.

Rip, Arie. 'Processes of Technological Innovation in Context – And Their Modulation.' In *Relational Practices, Participative Organizing*, edited by Chris Steyaert and Bart Van Looy, 199–217. Bingley: Emerald, 2010.

Russell, Stuart. *Human Compatible: Ai and the Problem of Control*. New York: Penguin, 2019.

Santoni de Sio, Filippo, and Giulio Mecacci. 'Four Responsibility Gaps with Artificial Intelligence: Why They Matter and How to Address Them.' *Philosophy & Technology* 34, no. 4 (14 December 2021): 1057–84. https://doi.org/10.1007/s13347-021-00450-x.

Santoni de Sio, Filippo, Giulio Mecacci, Simeon Calvert, Daniel Heikoop, Marjan Hagenzieker, and Bart van Arem. 'Realising Meaningful Human Control Over Automated Driving Systems: A Multidisciplinary Approach.' *Minds and Machines* (28 July 2022). https://doi.org/10.1007/s11023-022-09608-8.

Von Schomberg, René. 'Towards Responsible Research and Innovation in the Information and Communication Technologies and Security Technologies Fields.' Luxemburg: Publication Office of the European Union, 2011. https://doi.org/10.2777/58723.

Steele, C. M. 'A Threat in the Air. How Stereotypes Shape Intellectual Identity and Performance.' *The American Psychologist* 52, no. 6 (June 1997): 613–29. https://doi.org/10.1037//0003-066x.52.6.613.

Straus, Erwin W. 'The Upright Posture.' *The Psychiatric Quarterly* 26, no. 1 (1 January 1952): 529–61. https://doi.org/10.1007/BF01568490.

Van de Poel, Ibo. 'An Ethical Framework for Evaluating Experimental Technology.' *Science and Engineering Ethics* 22, no. 3 (1 June 2016): 667–86. https://doi.org/10.1007/S11948-015-9724-3/TABLES/3.

Van den Hoven, Jeroen. 'Value Sensitive Design and Responsible Innovation.' In *Responsible Innovation*, edited by Richard Owen, John Bessant, and Maggy Heintz, 75–83. Chichester: John Wiley & Sons, Ltd, 2013. https://doi.org/10.1002/9781118551424.ch4.

Van Oudheusden, Michiel. 'Where Are the Politics in Responsible Innovation? European Governance, Technology Assessments, and Beyond.' *Journal of Responsible Innovation* 1, no. 1 (2 January 2014): 67–86. https://doi.org/10.1080/23299460.2014.882097.

Vincent, James. 'Gender and Racial Bias Found in Amazon's Facial Recognition Technology.' *The Verge*, 25 January 2019. https://www.theverge.com/2019/1/25/18197137/amazon-rekognition-facial-recognition-bias-race-gender.

Young, Iris Marion. 'Throwing Like a Girl: A Phenomenology of Feminine Body Comportment Motility and Spatiality.' *Human Studies* 3 (1980): 137–56.

Part 4

Design for democracy

10 Design for democracy

Deliberation and experimentation

10.1 Digital technologies and democracy

The invention of the internet was acclaimed by many as one that could revolutionise global politics, by allowing for an unprecedented distribution of knowledge, a broader participation of people in public politics, and a great potential of breaking elite power and keeping governments under stricter control. Forty years later, the perspective has dramatically changed. While the potential to achieve these goals has not disappeared, it is far from having been realised. The massive increase in the amount of information available to the public has also created an increase in false or unreliable information. Public participation in social and political debates has widened, but the quality of democratic deliberation has not always improved. The internet has sometimes allowed to organise forms of popular resistance or counter-power, but it has also allowed political and economic elites to strengthen their power both in democratic systems – think of the outstanding case of Donald Trump in the US – and in non-democratic regimes, where the internet is under the direct control of the government. As Evgeni Morozov claimed already in 2011 in his book *The Net Delusion: How Not to Liberate the World*, the idea that the internet alone would make the world more democratic is a naïve illusion.[1] Technology tends to reflect the structure of social and political power in which it is realised, and if the social and political sources of non-democratic power are not challenged at their roots, these powers will not be shaken but rather supported by technological developments.

A strong awareness of the importance of structures of social and political power embedded in technology is increasingly present also in debates on the ethics of artificial intelligence (henceforth AI). In fact, there is a widespread call to focus more on the *politics* than on the *ethics* of AI. The call is summarised in the motto: 'Don't ask if artificial intelligence is good or fair, ask how it shifts power.'[2] This phrase expresses the concern that ethical reflections about the goodness or badness of the technology miss the point, as they do not sufficiently address the broader social and

DOI: 10.4324/9781003303244-14

political factors that may make technology development and use politically problematic. Discussions on the negative ethical impact of AI, and proposals to design 'AI for good'[3] or 'fair, accountable, transparent AI'[4] risk to become a purely academic exercise in the pejorative sense of the term, if technological development remains dominated by the principles of big profit-driven, unaccountable multinational companies[5] or non-democratic governments.[6] AI Ethics might become not only irrelevant but even harmful when ethical statements and standards are used by companies or political actors to increase their reputation while escaping harder regulations or leaving the deeper issues unaddressed. This is the phenomenon of so-called 'ethics washing.'[7]

Addressing this concern, as I see it, requires not replacing AI ethics with AI politics, but rather integrating more AI politics into AI ethics.[8] In fact, I have already tried to do this in previous chapters of this book. I have analysed freedom at work through the lens of theories of justice and the neo-republican idea of freedom as non-domination (Chapters 2 and 4), I have explained how internet users are increasingly dominated and exploited by big tech companies (Chapter 5), I have proposed to consider value-pluralism in the development and deployment of technology for professional use (Chapter 6), and I have proposed a concept of Meaningful Human Control over AI that reflects principles of individual moral responsibility as well as the social components of moral agency (Chapters 7–9). In this and the next, last two chapters of this book, I will address the relationship between AI and democratic freedom.

The reflection on democratic freedom and digital technologies is relatively new in the so-called analytic political philosophy, that is, roughly, the English-speaking one coming from the liberal-democratic tradition. This connection is much more established, as seen in Chapter 5, in the Marxist tradition.[9] It is also present in the field of research in philosophy and social science called 'surveillance studies,' strongly inspired by the work of French philosophers Michel Foucault and Gilles Deleuze.[10] According to Foucault, governmental power in modern societies operates and perpetuates itself more through automated social mechanisms than through vertical imposition. Norms and practices become embedded in social practices and institutions like schools and factories and are interiorised by people, perpetuated even in the absence of direct interventions by a government. The iconic representation of this modern form of power is the panopticon, the circular prison where inmates are constantly visible from a central point, while they do not see who, if anybody, is sitting in the central tower of control. In the panopticon, Foucault argues, power becomes automated, because people comply with norms independently from the intervention of a repressive authority.[11] Many consider Foucault's reading of power precious to understand how modern technologies have partially replaced the direct governmental power in ensuring people's compliance with social and political norms of the dominant groups.[12] Another source of inspiration for surveillance studies is Deleuze's idea of the 'societies of control,' which explicitly connects Foucault's conception of power with the pervasive character of invisible data-driven corporate power.[13]

Political philosophers Hoye and Mohegan claimed that surveillance studies influenced by Foucault and Deleuze have the great merit of exposing structural components of technology-mediated power and showing the limits of liberal-democratic approaches still focused on more traditional reading of (governmental) power and not sufficiently aware of the role of technology in perpetuating social norms and practices. At the same time, they criticise surveillance studies for not offering sufficient positive directions to address the issue of how to promote political freedom in the age of AI.[14]

10.2 Design for democracy

This criticism about the lack of a more constructive perspective may be extended to the literature on the political philosophy of AI. Most of this literature has focused on how AI may support negative phenomena such as algorithmic manipulation, biases, discrimination, perpetuation of old power structures, contamination of the public deliberation, surveillance, and negative environmental impact. In other words, how AI can damage democracy, rights, freedom, and equality.[15] A less-explored topic is how technology could be designed to promote and support democratic principles.[16]

This chapter and the next two try to fill this gap and explore the specific theme of 'designing for democratic freedom' or more concisely: 'design for democracy.'

The role of design in affecting or rather promoting democracy can hardly be underestimated. The more democratic participation becomes mediated by digital technologies, the more this participation can be consciously designed for.[17] Research in social science, computer science, and philosophy has already extensively studied and discussed the way in which, for instance, the current design of internet platforms may support systematic disinformation through the spreading of fake news or the polarisation of the political debate via the creation of so-called 'echo chambers'[18] and 'filter bubbles'[19] (people being exposed only to the kind of information they already have and prefer, due to the personalisation of internet content). Much less efforts have been devoted at defining some principles for the design of systems that may promote a healthier democratic participation.

In this chapter I will focus on the so-called deliberative democracy in the liberal and rationalistic tradition of Rawls and Habermas and democracy as experimentation and inquiry in the pragmatistic tradition of Dewey and Jane Addams. I have chosen these two approaches because, as I will argue, they nicely match two existing and well-developed 'democratic' design approaches: Value-Sensitive Design and Critical/Social Design. One claim of this chapter is that understanding the strengths and limitations of philosophical theories can help understand the strengths and limitations of what I present as their corresponding design theories, and vice versa. The choice of these two political theories and design approaches is based on their large relevance in their respective fields as well as the interesting matching they reveal.

In the next chapters I will explore four possible further themes in design for democracy: in Chapter 11, design for aesthetic democratic engagement (inspired by the philosophical theories of 'difference democracy' and 'adversarial democracy') and design for anti-power (inspired by the idea of 'anti-corruption' as key to democratic freedom); in Chapter 12, post-colonial design (inspired by decolonial studies) and more-than-human design (inspired by environmental philosophy). This is just an initial, incomplete, and not-exhaustive presentation of possible approaches to designing for democracy. More democratic theories and more design approaches could be considered and more interesting connections could be identified and presented. Future research will have to deepen and broaden the exploration of these and other avenues for designing for democracy.

Before moving on, two clarifications are in order. First, by looking at possible positive deployments and future opportunities of AI for democracy, I do not intend to minimise the current risks and threats posed by big-tech- and authoritarian governments- driven AI to democratic freedom. I just believe that the two lines of research – exposing risks and threats and exploring opportunities – must be pursued in tandem. The current political and economic picture may not leave much space for such more positive developments, but how can we expect a change to happen if we are not even clear about how an alternative may look like? Imagining and defining a desirable future is a necessary – if certainly not sufficient – condition to make that future happen. As a matter of fact, as I will show, some interesting experiments of designing for democracy already exist. Making their political potential more visible is another crucial step to support and expand them.

However, one could push back and say that, if we also take seriously the environmental and human impact of the current development and use of AI,[20] we should just conclude that AI should not be produced anymore, or its production should at least be dramatically downscaled. And that this should be the most urgent cultural and political campaign to support now. I agree that current modes of development and use of AI technologies, driven by big multinational companies for commercial purposes and fed by the massive extraction of natural resources, human labour, and data, are problematic, should be criticised, and where possible regulated. I have expressed my own concerns about the dominating and exploitative character of present-day development and use of AI technologies by (big) companies in previous chapters, and I will say more in the next ones, with particular attention to its colonial and extractive dimensions.[21] But AI and digital technologies more generally, in one form or another, are here to stay, and many applications promise to be beneficial. So, the question remains: if anyone wants to build a democracy-promoting AI, what would this look like? What roles would people and technology have? And, to start with, what should be the role of actors that are less dependent on private companies or authoritarian governments in this enterprise, such as civil society, academic and independent designers, or (democratic) governmental agencies?

While recognising the merits of more legal and political approaches, grounded in regulations and more traditional political initiatives and interventions, especially in limiting the power of tech companies, I will present a complementary perspective, grounded in design. As I have already claimed in previous chapters, and will reiterate in the next two, design – broadly construed – is aimed not only at producing new things but also at supporting a critical, ethical, and political approach to technology. When performed in this spirit and in close collaboration with the people it must serve, design can be a powerful means to imagine, experiment, and realise a more democratic future.

10.3 Two projects in design for democracy

Consider the following two existing projects, exemplifying two different ideas of 'design for democracy' and of democratic design.

10.3.1 e-Democracia

In 2009, the Chamber of Deputies, the lower house in Brazil's national congress, set up a portal called e-Democracia. This was a pilot project to engage citizens in the legislative process. Led by an office called 'Strategic Management and Projects,' it followed the recommendations of a multidisciplinary team of parliamentary staff, consultants, and digital experts. After mapping existing legislative experiments from around the world, the team suggested to 'adapt the mechanism of virtual communities of practice to the formulation of laws to engage a broader segment of society in debates of national legislative issues.'[22] In other words, to include more people in the legislative process, by using more online social communication tools. The project had some ups and downs, including getting close to being cancelled in 2013 and then eventually surviving, also following a public protest against the planned cancellation.

The e-Democracia website is organised into three main areas: virtual communities on thematic areas, 'free space,' and Wikilegis, a tool for drafting bills collaboratively. The platform has been used to discuss a wide number of critical draft bills, including a climate change policy, a new Youth Statute, a bill on the Amazon and a space policy.[23]

The experiment has been successful in two ways. First, it has led to significant legislative results. The Youth Statute Bill crowdsourced 30% of its final text from young people across the country, and the Internet Civil Rights Bill received 374 individual contributions on the Wikilegis platform, many of which were explicitly referred to by the rapporteur in the Chamber, and adopted in the final bill.[24]

The initiative also generated interesting spin-off design experiments in connecting people around technological projects to improve the functioning of the platform. In 2013, designers, developers, parliamentary staff, and representatives joined a 'hackathon' to develop new apps for computers and mobile devices to further improve transparency and the public's understanding of the legislative process. The event was such a success that a permanent participative hackerspace – LabHacker21 – was created within the Chamber to act as an innovation lab and to forge links between parliamentarians, designers and developers, and civil society actors.[25]

The Brazilian e-Democracia project is not unique in its genre, and many parliaments, municipalities, and political parties around the world – from Iceland to Taiwan – have recently introduced digital platforms to increase and expand the participation of citizens in political processes.[26]

10.3.2 Careful Coding

In 2017, Georgia Tech designer Carl DiSalvo and colleagues turned to Atlanta's Black neighbourhood English Avenue. They team up with Block by Block, an existing data collection initiative run by a local resident with the help of local students. The goal of the initiative was to report the so-called 'code violations' - instances of neglect and poor maintenance of the built environment, which could cause collective problems such as deterioration of public spaces or rat infestations- while at the same time understanding (and sympathising with) the circumstances that cause these violations such as poverty and exclusion, giving local people agency to determine what is acted upon, not burdening people with excessive 'punishment' for these violations but if possible helping them to address their issues. DiSalvo's initiative to support Block by Block was called Careful Coding.[27]

The project presented some technical challenges, such as in coupling pictures to digital maps (mapping). But the biggest challenge was one of 'institutional design.' How to collaborate with the municipality? How to combine existing data and how to define the roles of different actors?

DiSalvo's team managed to use their technical expertise to facilitate the creation of new formal and informal interactions between the municipality and the local association Block by Block. Block by Block was eventually recognised by the municipality of Atlanta as a trustworthy collaborator whose data could be used and integrated into the city activities while also maintaining its autonomy and power in deciding how to collect data in a way that cares for the conditions of the local communities.[28]

10.4 Design for democracy and democratising design

Both the e-Democracia and Careful Coding projects aim at enforcing democratic participation through the design of digital tools and can be described under the header of 'design for democracy.' But they look different. e-Democracia is a big project, run by the government and designers hired by them, that aims at engaging Brazilian citizens in deliberation over legislative and policy issues via new digital tools. Careful Coding is a project run by a team of academic designers who used their technical and design skills to support and empower a local initiative in data collection for social purposes.

These two projects embody and represent two different views of democracy and how design can support its development. They also represent two different views of what design is and how it can become more democratic. These two views of democracy and design are not incompatible, and indeed, as I will argue, we may rather need both, or a combination of the two, to reinforce democracy by design. It is therefore important to clearly identify their respective strengths and weaknesses. This is the main goal of this chapter.

In a nutshell, so-called e-government and open government projects like e-Democracia see people participating in the democratic process in a way that looks very close to the theory of deliberative democracy, where institutional set-ups should be designed to promote a fruitful and rational dialogue between people over issues of public interest. If designed correctly, the digital space may be the twenty-first-century's *agora*, the public space where citizens freely discuss and decide on political issues. On the other hand, projects like DiSalvo's Careful Coding are inspired by a pragmatistic reading of democratic participation, where, in a sort of reversed process, it is through people's active participation in experimental projects addressing specific societal issues that new institutional set-ups are devised and created.

When it comes to design, e-Democracia and Careful Coding reflect two different approaches. In e-Democracia, the design process is mainly in the hands of policymakers and designers to realise a platform that promotes good deliberation practices. Even when the participatory design platform LabHacker was launched as a spin-off of e-Democracia, designers and citizens were mainly asked to contribute to the improvement of the platform and the operationalisation of some relevant principles and values. In design theory, such an approach can be connected to the ideas of Value-Sensitive Design[29] and Scandinavian Participatory Design.[30]

In Careful Coding, the professional designers contribute to an existing data collection initiative run by residents of a Black neighbourhood. Designers act as data and institutional intermediaries between the local initiative and the municipalities and help imagining, investigating, and experimenting new institutional set-ups to give the local initiative more political power and agency. Such an approach derives from two other trends in design theory: Critical (or Speculative)[31] and Social (or Participatory) Design.

As I will argue, the strengths and limits of the deliberative democracy approach are, to some extent, also those of Value-Sensitive Design, and vice versa. And the same holds for the pragmatist approach to democracy and the Critical and Social approach to design.

The next two sections dive into the two philosophical theories of democracy – deliberative and pragmatist – and how they are reflected, respectively, in the e-Democracia and the Careful Coding projects. The successive two sections present the two different approaches to design exemplified by the two projects and reflect on their potential to promote democracy in the design process. The final section recapitulates and concludes.

10.5 Two theories of democracy

10.5.1 Democracy as rational deliberation

In very broad terms, supporters of deliberative democracy aim to create or reinforce institutional spaces where the exchanges between people tend to promote a fruitful, constructive dialogue between reasonable positions over relevant issues, as opposed to a mere clash of interests or powers, or opposition of pre-rational perceptions. In twentieth-century Western political philosophy, this deliberation space is often referred to as the domain of the 'public sphere' (following Jürgen Habermas)[32] or 'public reason' (following John Rawls).[33] This is an ideal, normatively defined space that, while never fully realising in existing societies, should be promoted and sustained via procedural conditions and institutional rules. According to this approach to deliberative democracy, while in a pluralistic society people can and will disagree on many or potentially all subject matters, they should at least agree on respecting the basic rules of a well-ordered discussion. These rules are considered not a limit to political freedom but rather its fundamental pre-condition. Kant famously wrote that the dove may perceive the air as an impediment to its flight, even though it is rather what makes flight possible. All rules are somehow constraints to freedom, but some constraints are necessary to define and sustain it. The rules of a chess game limit the freedom of players, but they also make the game possible. No rules, no game. In philosophical jargon, these are sometimes called 'constitutive rules.'[34] And indeed, Rawls and Habermas, two of the main inspirators of present-day deliberative democracy theory in Western philosophy, do somehow share a Kantian approach to the rules of the 'public spheres' or 'public reason.' In their view, while these rules may sometimes be perceived as a limitation of expression by participants in a political debate, it is thanks to these rules that a free political deliberation, as opposed to a mere struggle of opinions, or clash of emotional reactions, can happen.

Deliberative democracy is also thought to bring, broadly speaking, some additional, practical benefits.

First, it arguably better represents the people's will (in a pluralistic world). The idea is that democratic participation will promote a better responsiveness of the decision-making process to people's judgements and preferences. By having people involved in decision-making, there are better chances that the decision-making will reflect their will. This responsiveness is good for people because, as Tim Scanlon once said, if you are allowed to order at the restaurant, you are more likely to eat what you want or what you think is good.[35] But it also improves the legitimacy of government decisions.

Second, it may lead to better collective decisions, where better is assessed through independent, objective standards and not defined as being more representative of people's will. This is, depending on the approaches because of the tendency of free criticism and deliberation to identify problems and push for governmental responses more quickly and effectively or because of the tendency of democratic practices to create a faster and better circulation of information. For instance, Amartya Sen famously claimed that 'there has never been a famine in a functioning multiparty democracy.'[36] The last big famine in India happened in Bengal in 1947, just before the end of British rule. The spreading of the famine was caused by a series of bad governmental decisions which, according to Sen, 'A democratic system with public criticism and parliamentary pressure would not have allowed.'[37] Similarly, the tragic famine in China (1958–1961) could have been stopped earlier if the government had access to more precise information about the reserves of food present in the country farms. But local party officials did not have this information or they did not share it with the government, not to lose credit.[38] This, Sen argues, would not happen in a free democracy where opposition parties and free press would acquire and circulate information.

Sen (and others) also claim that public deliberation is good for people in yet another way, because it improves their self-understanding and capacity for reflexivity. By engaging in discussions with others one ends up refining their own arguments and positions and clarifying their views and preferences. Public deliberation, as it were, helps people build up their reflexive capacity and moral autonomy.[39]

So, according to their supporters, democratic deliberative procedures are not only the best way to make sure people will get what they want or what they think is good; in other words, they are the best way to respect people's free will (and thereby preventing abuses of power and oppression by the government); they would also have an epistemological superiority, by being the best way to manage and aggregate big amount of information coming from different people and places. And they would even be beneficial to people by helping them build up better beliefs, overcome their cognitive limitations, and become more rational and mature persons.

In 2017, NESTA, a UK-based think tank, collected and analysed information about many existing e-government and open government initiatives around the world.[40] The projects covered in the report are run by Parliaments, municipal governments, or political parties and invite citizens to use new platforms to engage in deliberations, proposals, and decision-making.[41]

These initiatives aimed to design new digital spaces to create or reinforce a 'public sphere' and the practice of 'public reason,' with all its potential benefits. To what extent did they succeed in their goal of increasing the people's participation in democratic participation? And what do these experiments say about the democratic potential of a digital public sphere?[42]

All of these projects started from a shared consideration: there is a big gap between the daily life of citizens and the way politics and democracy are carried out.[43] While almost every sphere of life – work, leisure, and social relationships – has been dramatically transformed by new technologies, 'representative democracy bureaucratic administration, occasional elections, and rule by elected or appointed "experts" – remains largely unchanged since it was invented in the 20th century.'[44] This is arguably one key factor contributing to declining interest, trust, and confidence in democratic institutions. It is time to bring our democratic procedure to the twenty-first century – so the argument goes – and technological updates are a key step in the process.

The NESTA report confirms that citizen participation, when properly incentivised and structured, has great potential to improve the quality of public deliberation and decision-making, especially at the local level. 'Local residents better understand their needs and those of their communities but also because the platform ensures that poor ideas get sifted out by the wisdom of the crowd.'[45] In Reykyavik, one of the early successes was an idea put forward by a nine-year-old girl requesting more school field trips.[46]

Also, citizens' participation in some cases also contributed to improving the technical platforms themselves. See the success of the LabHacker as a spin-off of the e-Democracia project in Brazil (more on this below).

Sometimes, in line with the wishes of supporters of deliberative democracy, citizen participation has done more than just contributing to reaching better decisions. It has also allowed citizens to improve their reflection and awareness. For instance, the platform Pol.is, as used by the vTaiwan project, invites citizens to list their pro and con argument on given proposals, without giving them the opportunity to criticise or directly engage with the points raised by others (as happens on mainstream social networks). This has arguably allowed people to focus on the content of their reflection, rather than on attacking and defending from others. The reflection was enhanced by seeing all arguments listed next to one another. This also helped individuals to acquire a better awareness of their exact position on a given topic.[47]

Some relevant lessons learnt are also presented in the NESTA report. Giving people real power is key to the success of digital democracy projects. This can be done by making votes binding as it happens in the Decide Madrid platform, or by giving people the power to decide on significant budgets, as in the case of the participatory budgeting processes in the projects Madame Mayor I Have An Idea (Paris), Better Reykjavik, and Decide Madrid.[48] Also, it is important to engage citizens as early as possible, to give them the chance to set the agenda and frame the problem. The vTaiwan process is designed in such a way that the early stages are genuinely open-ended.[49] This is in a way not surprising: the point of democratic procedures is giving people the power to make their will matter, not just to speak their mind or provide advice to the government.

Ironically, the biggest limitation of these experiments in digital democracy has so far been participation. Even if relatively little data are available about the profile of the participants, as many platforms avoid requesting filling in long registration forms to lower the barrier of participation, the available data suggest that 'participation is skewed to those who are already politically active and towards well-educated, young men in urban areas.'[50]

Another big challenge has been the processing of data. Many projects decided to reduce the filtering and moderation of the debate to not make people feel censored or silenced, but this has often made it more difficult for platforms to effectively manage the information flow and keep track and organise all the different proposals and arguments.

Optimists may see these limitations as the exact space where good policies and better technological design can make a difference. People are simply not accustomed to these new forms of participation. Digital literacy should be improved and governments should learn how to invite people to participate more. Also, the technology used in the digital platforms should be improved. To name a few, AI techniques may help filter the big amount of data produced on these platforms and help clustering proposals into groups. As it has been proposed in relation to the platform economy, 'the technological heart' of big social network platforms can be 'cloned' and put at the service of democracy.[51] Here the design challenge would be that of designing platforms that entice people to participate more, without falling in the trap of sacrificing the quality of the discussion for the sake of increasing engagement, as it often happens on social network platforms run by big technological companies. Finally, other digital technologies like Virtual Reality may contribute to make the deliberation process more realistic – people may be immersed in a virtual room with other participants – and allowing for a better capturing of people's emotions, for instance, by capturing facial expressions.[52]

However, one may also take a more sceptical look at these projects. As successful as they may be (or become) in engaging people in public deliberation and public decision-making, they may be missing out on something important about democracy. The idea of democratic participation in which they are grounded may be itself insufficient, so that no matter how much new policies and technologies may help realise that idea, democracy would still not be fully realised. This is what the supporters of the idea of democracy as experimentation and inquiry would argue. The next section elaborates.

10.5.2 Democracy as experimentation and inquiry

Jane Addams was a social scientist and activist affiliated to the Department of Sociology at the University of Chicago. In 1889, together with Ellen Gates Starr, she established a boarding house for immigrants in Chicago, called Hull-House. Hull-House would offer many spaces for shared activities such as classes in bookbinding and crafting homemade soft drinks. At some point, the concern had emerged in the community that striking women might be more at risk of being evicted from Hull-House, due to their more precarious financial situation; if they ever lost their job, they could not afford the board anymore. Addams recalls that during a meeting of working girls at Hull-House, after sharing the hardships experienced by those who were at risk of eviction, one of them exclaimed: 'Wouldn't it be fine if we had a boarding club of our own, and then we could stand by each other in a time like this?' After that meeting, they read together a little book on 'Cooperation,' started discussing opportunities and problems of their project, and after a couple of years, managed to rent two apartments near Hull-House for their cooperative hostel.[53]

At the Department of Sociology at the University of Chicago, Jane Addams was a close collaborator of John Dewey, the philosopher considered as one of the fathers of the philosophical tradition known as pragmatism. Her projects were explicitly rooted in this approach. According to Dewey, at the core of democratic life is not, as deliberative democrats hold, abstract thinking and rational deliberation. The core of democratic life is the so-called 'social experiments.'[54]

According to this idea, democracies should not just realise and enforce existing deliberative practices and institutions. Like a scientific enterprise, democracy must also continuously reflect on the validity of its methods of representations, putting itself into question, and experiment with new methods. One may compare this with the relationship between science and engineering. According to one simplistic and largely inaccurate reading, science develops the abstract principles that are then applied by engineering. In fact, the relationship runs both ways. It is often thanks to some design intuitions – for instance, about an experimental set-up – or technological discoveries – think of Galileo's telescope – that scientific breakthroughs and discoveries and theories are developed.

Similarly, the norms and principles of (democratic) life are rarely the result of only abstract reflection and deliberation. They are more often the result of practical experiments run by people who get together because of individual concerns which, in the beginning, may not even be explicitly perceived as a shared, political issue. This experimentation is, in the pragmatist tradition, a way of doing 'democratic inquiry.'[55] Because it is through these practical enterprises that new values, principles, and norms can eventually emerge and being defined. Also, it is through these 'cooperative investigations' rather than just by being exposed to each other's abstract arguments that individuals become conscious political actors, aware of their (shared) concerns, and able to actively shape the (institutional) environment in which they live.

There are important differences between the pragmatist approach and deliberative democracy theory. First, according to the pragmatist approach: public reason is not only discursive, abstract, and theoretical but is rather 'directed towards the material control of actions and their consequences' and 'carried out by communities of enquiry and experimentation.' Second, the subjects of public reason do not pre-exist but are rather shaped and constituted through the exploration and experimentation of new social arrangements. Third, and relatedly, the ultimate goal of pragmatist social experiments and democratic inquiry is not only to create a space for discussion about what is good or right, but rather, based on a drive for the good and the right, to change the (social) environment: to 'reshape the material equipment, established knowledge, collective habits, professional careers, and institutional agencies.'[56] While deliberative democrats aim at reinforcing and expanding an ideal 'public sphere' where 'public reason' can be practised, pragmatists aim at encouraging experimental set-ups where, through the experience of collective and cooperative actions, groups of individuals can constitute themselves into 'publics' (plural), that is, as a belonging to a social group with a political awareness. In other words, people not only discover what they really believe by engaging in a rational exchange with others but also discover what they care about and determine who they are through action and interaction in situated contexts. Local experience and imagination are at least as important as discursive deliberation.

At the core of the pragmatist tradition, then, lies a strong push towards democratic innovation and, we may say, design. '[P]ublics are not only dependent on their environment, they also create it.' The (pragmatist) public 'can never be fitted over what existed previously' ... 'it constantly rearranges living environments and redirects the ways in which people see, say and do.'[57]

But to endure and persist through time, these experimental rearrangements should stabilise and be institutionalised. 'Institutions capitalise on [...] creative intelligence, work to transmit and reproduce their results, set new cultural habits, and incubate new professional careers.' In Dewey and Addams' time, 'these institutions were schools, universities, theatres and libraries, social centres, political clubs, alternative press outlets or adult education centres; today, these institutions include digital platforms, whistle-blowing systems, databases, NGOs, parallel markets or exchange networks.'[58]

The Careful Coding project presented in the opening of the chapter is included, together with other projects, in Carl DiSalvo's 2022 book significantly titled *Design as Democratic Inquiry*. DiSalvo explicitly endorses the pragmatist approach to democracy and explains to what extent projects like Careful Coding can be seen as what he calls 'design experiments in civics.'[59] As he explains, the word 'civic' is used because these projects are about forms of democracy 'in the small.' Working in the small, as he noted, is certainly something at odds with contemporary trends in policy and design:

> Global design consultancies regularly work with national governments and civil society organizations to develop programs, strategies, and policies. During an era of 'big data' – when everything must occur 'in the large' – the work of design and democracy can also be more modest; the context of the local and mundane is also important.[60]

DiSalvo's projects are 'design experiments' because, in a pragmatist fashion, design does not necessarily aim at a predetermined final product whose functionalities may address immediate material needs or solve problems but rather employ imaginative making to produce 'artifacts, systems, and services that are both sites and modes of inquiry.'[61] Rather than problem-solving, DiSalvo argues, his experiments are the first sources of problem-making, in that they 'identify and articulate problems that should be addressed, perhaps by people other than designers.'[62]

In the Careful Coding project, DiSalvo and his colleagues from Georgia Tech identified a challenge that was at the same time technical and institutional. Both the municipality of Atlanta and the local initiative Block by Block were collecting data about the so-called 'code violations' in the neighbourhood of English Avenue in Atlanta – situations of neglect and poor maintenance of the built environment which could cause collective problems such as deterioration of the public spaces or rat infestations.

The general goal of the Careful Coding project was to integrate the municipality's approach on the one hand and Block to Block's on the other. The municipality's approach was mainly oriented at collecting as much information as possible on the state of the neighbourhood and then getting people to fix violations through formal interventions: fines and legal actions. Existing civic tech platforms in Atlanta, like 311 or SeeClickFix, would indeed allow citizens to route information directly to the city government or a third-party Intermediary, and these would then act accordingly.[63] Block to Block, while sharing the ultimate goal of keeping the neighbourhood clean and well-maintained, also wanted to take a more nuanced and socially sensitive approach, one that would also pay attention to the circumstances that cause neglect, such as poverty and exclusion, and not burdening people with a low income with unbearable legal procedures and fines.

The technical challenge of the Careful Coding project was therefore helping Les from Block by Block to improve their capacity for data collection (in the form of geo-localised photos and texts), where improvement meant not only more data or more accurate data but also data that could be associated and integrated with the database of the city. At the technical level, the design collaboration between DiSalvo and Les consisted in a virtuous exchange of know-how, data and resources. While Les contributed with his knowledge of the neighbourhood and the many data already acquired by his organisation, DiSalvo and colleagues provided him the Georgia Tech know-how of working with data as well as material resources like access to digital tablets and Wi-Fi hotspots.

However, the integration of the data sets of Block by Block and the city was not only a technical challenge but was first and foremost an institutional one. The institutional challenge consisted in bringing the city to integrate Les' data into their operation while at the same time leaving Les the possibility to maintain his socially sensitive approach to data collection and his capacity to influence the city's decision on activating legal proceeding against code violators.

DiSalvo took this shift of role from data intermediary to institutional intermediary, and from designing tools to staging encounters with city government workers, to be at the core of his 'design experiment in civics.' He presents the institutional mediation following the technical work on the ground in some detail.

Les from Block by Block proposed that only the Georgia Tech folks would approach the Code Enforcement Section of the city of Atlanta; he rightfully thought that their status would grant them more credibility. After making contact via email, DiSalvo and colleagues were summoned at the office of the Code Enforcement Section. In a first brief, not very comfortable, meeting, DiSalvo's group managed to overcome the officers' scepticism about the usefulness of their project and explained their willingness to help by providing the data they had acquired on the ground. They were eventually granted another appointment, this time with the officer responsible for Les' neighbourhood, and on the ground.

In this second meeting, in which also Les participated, the conversations were 'open and at times unsettling.' And, DiSalvo remarks, it could not be otherwise, as 'we were exploring our own and each other's boundaries, our own and each other's obligations and desires, listening for opportunities to shift into new configurations; and also, in some moments, guarding existing configurations against any such shift.' In other words, in those open encounters and conversations, the neighbourhood activist and the city officer, with the crucial mediation of the designers, were experimentally defining a possible new institutional set-up for code violation enforcement, a set-up that would include the use of new data and technical tools but most importantly: new combination of formal and informal rules, and new political identities and powers for the agents involved.

In a series of exchanges that, at least in DiSalvo reconstruction, climaxed in dialogues worthy of a Netflix show, the officer explained to Les some formal rules about the use of data – for instance, Les learnt that pictures taken from his private phone were useless as they could not be used by the city as evidence in a case in court. The officer also provided him with an important informal insight: to avoid delays in the reporting, he could use her private phone number; in practice, she offered him an informal agreement which granted him faster and privileged access to the communication with the city.

Les then addressed one crucial and sensitive issue: his commitment to not report some violations, to spare some people the legal and financial ordeal that could follow a report:

> He tells her that sometimes he just wants to look around and keep track of things before he tells anyone. She takes a pause, just looking at us. Then she responds, 'That's good, because once you send it to

> me, I've got to do something about it. So if you don't want me to do something about it, don't tell me. If it's on my plate, I gotta serve it.'

So, Les was *de facto* recognised the option to take some distance from official policy and procedures without compromising his collaboration with the city.

Then, suddenly also the designer's ambiguous role and 'intersectional subjectivity' – as an academic researcher, a university employee, and a political activist – was called out by the officer.

> Then she looks at me. 'You gotta decide whose side you're on. Are you on the side of the university or him?' pointing to Les.
>
> 'I'm on his side,' I stammer, taken aback by her directness.
>
> 'Uh-huh,' she says, with doubt. 'Well, you better be ready for the call.'
>
> 'What call?' I ask.
>
> She says, 'Here's what's going to happen. You're going to report something and it's going to come to me and then it's going to turn out that the house I just put a sticker on because you helped him call it in belongs to your president's auntie, and now all of a sudden he's in a fuss. And I don't care about your president or his auntie, and so then you know what's going to happen – she's going to get a citation and then he's going to call you. That call.
>
> 'I'm okay with that, I'll take that call,' I reply.
>
> 'Uh-huh,' she says again. 'You have to think about that and figure it out. Y'all are going to have to figure it out, how you want to make this work.''

The final line nicely captures the spirit of pragmatistic design experiments in civics.

Projects like Careful Coding are projects 'in the small,' in the sense that they address one specific issue in one specific community, in this case the management of code violations in one neighbourhood of Atlanta. To be impactful, these projects should produce technical and institutional effects that go beyond the project, not necessarily in size but certainly beyond the time span of the single project. They must create not only new technical tools and procedures but also stable arrangements, in the form of new formal and informal rules and institutions, for instance, new ways of collaborating in the collection of data between the city and the local organisation. And in the form of new commitments and identities for the agents involved, for instance, a new role as a collaborator of the municipality for the local activist Les (and a new role as an activist for the designer DiSalvo).

Even though in such projects designers play a major social and institutional role by becoming mediators between social agents, and even sometimes activists in a process of social change, they still do not give up their role as 'makers' and experts in technology. They just reinterpret those roles in a different way. DiSalvo's team recognised that academic expertise in the design of processes and tools for data collection, for instance, played a key role in the success of Careful Coding. Without that expertise they would have not had the same access to and impact on the work of Block by Block and the City of Atlanta. Still, they used that expertise not only to improve the technology but also to facilitate a better use of the existing socio-technological expertise of the people involved.

As the story of Careful Coding clearly illustrates, socio-technical arrangements are never neutral, and indeed DiSalvo acted with a clear commitment to design for what he considered to be a more democratic set-up, one that would shift some agency and power from the city to the local organisation Block by Block. Concretely, this was achieved by allowing Les from Block by Block to retain control over what was shared, how, and with whom.

10.6 Two design approaches

Both the projects and approaches above show the importance of design practices for reinforcing democratic practices. So, in addition to asking which democratic theory designers should embrace, we should also ask: which design theory should democratic societies embrace? Or, in other words, what does it take to democratise the design process?

10.6.1 Value-Sensitive Design and (Scandinavian) Participatory Design

At a closer look, the ideals of Responsible Innovation and Value-Sensitive Design presented in Chapter 3 nicely reflect the philosophical ideal of democratic deliberation. According to deliberative democracy theory, the systematic involvement of citizens in public decision-making promises to increase the quality and legitimacy of the decisions (in addition to respecting people's equal right to participate). Similarly, the systematic involvement of so-called stakeholders in the design process promises to increase the quality of the technological product and legitimacy of the innovation process. Indeed, as seen also in the broader literature on Responsible Innovation, to which Value-Sensitive Design is connected, the involvement of stakeholders is meant to promote better responsiveness of technological development to the people's values and interests,[64] prevent unwanted outcomes by tapping into stakeholders knowledge,[65] and eventually allow for the creation of better technological products where better means technically better due to more information and expertise being involved in the design[66] as well as better as more representative of people's and society's will and needs.[67]

Some supporters of Value-Sensitive Design have proposed that a design process may even become a space for the creative solution of not only difficult technical challenges but also thorny societal issues.[68] Design processes often encounter value trade-offs, situations in which it seems impossible to increase the realisation of a certain value or parameter in a process or product without negatively affecting another desired value or parameter. In a standard example concerning digital technologies, one may reach a point where it is not possible to improve the security of a given area through surveillance technology like cameras without negatively affecting privacy, and vice versa. This stalemate may be partially overcome through Value-Sensitive Design. The systematic interaction of designers and stakeholders may allow them to better specify and operationalise the values of security and privacy in a way that new combinations of them become possible, and the trade-off can be loosened.[69] As a simple example, it may turn out that, for stakeholders, privacy does not necessarily require removing the cameras but only limiting their capacity to capture good-definition pictures of passers-by, and police or municipalities may find out that security can also be improved by adding cameras with low-definition capture. Also, a process where relevant stakeholders are involved in the design decision and are therefore aware of the functionalities of the cameras and the reasoning behind its design are more likely to gain stakeholders' trust once the cameras are implemented. In other words, trust is enhanced when not only privacy but also the *perception* of privacy is preserved.

A similar democratic spirit also animated another innovative design approach: Participatory Design, as originally developed in the Scandinavian countries. Originally, this meant that workers were invited to participate in the design of the technical systems they would use, to ensure that these tools would better fit their will and needs, as the users of these tools.[70] In more recent times, and beyond the workplace, the concepts of user-centred or human-centred design have promoted similar ideas.[71]

In the Brazilian e-Democracia project, we have seen how citizens' participation via the LabHacker platform has in some cases also contributed to improve the functionalities of technical platforms for e-government and creating new tools. The success of a hackathon prompted the creation of a permanent 'hackerspace' called LabHacker, a place where parliamentarians, designers, developers, and civil society actors can discuss, propose, and realise new digital tools to further improve transparency and understandability of the legislative process and to encourage people's participation.[72]

10.6.2 Critical and Social Design

As groundbreaking and innovative as they have been in showing the need for the democratisation of the design process, Value-Sensitive Design and Scandinavian Participatory Design have also been recently criticised for not being sufficiently democratic.

In *Design as Democratic Inquiry*, DiSalvo proposes to look at two other traditions in design studies: Critical/Speculative Design and Social/Participatory Design (beyond its original Scandinavian form).

'Critical Design' is a term coined by Anthony Dunne and Fiona Raby in the mid-1990s to describe a design practice that 'uses products to ask questions and raise issues in society and culture.'[73] It typically uses prototype products to prompt questions about possible and (un)desirable futures that technological development can lead to. In a way, it is not far from science fiction and other forms of art where the technical production is used to 'push the limits of the lived experience.'[74] It asks us not only what kind of technological products we as 'stakeholders' want or need but more broadly what kind of society we as people and citizens want to live in and what should the role of technology be in this society.

In line with Langdon Winner's idea of 'epistemological luddism' presented in Chapter 2 and the broader idea of critical theory in philosophy and sociology (most famously represented in twentieth-century Western philosophy by the Frankfurt school), the key idea of Critical Design is that social change should start by becoming fully aware of the forces and structures that shape and steer current society, politics, and economics, the impact and potential (negative) consequences of accepting these forces and structures. And most importantly, that other, more desirable, forms of life are possible. But these new forms of life may just be difficult to imagine. That's why critical social analysis, experience, and imagination are key to Critical and Speculative Design.[75]

However, to start realising social change, imagination and one-off design prototypes or artistic projects are not enough. To be politically effective and produce lasting consequences, imagination should be enacted. This is why DiSalvo and other design theorists like Laura Forlano propose to combine Critical Design with the tradition of Social and Participatory Design.[76]

In *Design Justice,* Sasha Costanza-Chock insists on the risk of an approach to stakeholder participation that ends up being 'extractive' rather than genuinely participative.[77] They also insist on the importance of collaborating with communities by recognising and valuing the expertise they already have, and expanding the expertise they need to further flourish, rather than seeing stakeholders as a source of precious knowledge and information to be exploited by designers for their own purposes or that of companies, governments, and others.

Binder and colleagues also problematise the concept itself of 'stakeholder involvement.' Stakes and stakeholders – they notice – 'were conceptualised in the realm of organisations whose activities are classified as "projects."' But as 'the focus of codesign research is shifting to broader societal issues and public concern, the rhetoric and assumptions about stakes and stakeholders are also being challenged.'[78]

And in a truly pragmatist fashion they claim:

> This kind of assumed a priori juxtaposition of groups defined by a more or less singular 'stake' is not very conducive to the tentative formation of new publics and issues. We should think of the design process not as much as a deliberation and negotiation between already formed groups, interests and opinions, but rather as the space where these groups, interests and opinions are formed.[79]

This is 'a fundamentally experimental attitude to the epistemology of the collaborative encounter.'[80]

Playing a bit with words, we can say that 'democratic deliberation' and 'stakeholder involvement' work well for *decision-making*, but democratic experimentation in design works better for *sense-making*.[81]

This is indeed the approach that inspired DiSalvo's pragmatist approach, where the process of design aims at articulating issues to let agency form around emerging publics. In those cases, design does not perform the agency, but rather prepares for a public to form and for some sort of civic engagement to be enacted.[82]

The role of the designer is quite different in these two approaches. In Value-Sensitive Design and Scandinavian Participatory Design, designers remain mostly experts in 'making,' who retain the role of coordinators of a complex process where they try to combine the knowledge, interest, and expertise of different stakeholders. They aim to create a product or process that embeds the best possible combination of these contributions and aspirations. In the democratic experiment approach, the designer is an expert in 'doing,' who facilitates the collaboration between societal actors who are already involved in the creation of their own products and processes. The designer is a socially engaged facilitator who creates a space where these activities and goals can converge, possibly enter in tension among each other, thereby prompting the emergence of new forms of social agency and power. They aim to create new forms of agency and power and to consolidate emerging publics and new institutional set-ups for these new publics to act in the democratic arena. The critical and imaginative push of Speculative Design is combined with the hands-on, socially situated work of Social and Participatory Design.[83]

Table 10.1 Summary of the two models of democracy and design presented in the chapter

Model of democracy	*Deliberative democracy*	*Experimental Democracy*
Case study	e-Democracia[a]	Careful Coding[b]
Philosophical references	Rawls,[c] Habermas[d]	Dewey,[e] Addams[f]
Main activity	Rational discussion in the 'public sphere'	Experimental creation of new forms of engagement
Role of people	Participants in the deliberation as citizens	Creators of new active groups around specific matters of concerns ('publics')
Design methodology	**Value-Sensitive Design, Scandinavian Participatory Design**	**Critical, Speculative, Participatory Design**
References	Batya Friedman,[g] Jeroen Van Den Hoven[i]	Carl Di Salvo,[h] Sasha Costanza-Chock[j]
Role of designer	As a technical expert	As expert and activist
Design activity	Creating digital spaces for deliberation	Supporting communities in developing their existing projects and skills
Role of design	Creating a new system that support rational deliberation	Facilitating democratic experimental engagement
General goal of design	Problem-solving	Problem-making, sense-making
Expected result of design	Better technical systems, better deliberations, more rational citizens	New forms of engagement, new 'publics'or 'civics', new norms and institutions
Typical dimension	In the large	In the small
Typical direction of action	Top-down	Bottom-up

a Simon et al., *Digital Democracy: The Tools Transforming Political Engagement*, 19–23.
b DiSalvo, 'Design as Democratic Inquiry,' 103–38.
c Rawls, 'The Idea of Public Reason Revisited.'
d Habermas, *The Structural Transformation of the Public Sphere.*
e Dewey, 'Challenge to Liberal Thought.'
f Addams, *Twenty Years at Hull-House with Autobiographical Notes.*
g Friedman, 'Value-Sensitive Design.'
h Van den Hoven, 'Value Sensitive Design and Responsible Innovation.'
i DiSalvo, 'Design as Democratic Inquiry.'
j Costanza-Chock, *Design Justice.*

10.7 The future of deliberation and experimentation

Table 10.1 summarises the main differences between the two approaches to democracy and design as presented in this chapter. This is an abstraction, and the two models are 'ideal-types' which may not fully reflect existing projects.

Which way to go? Which approach to prefer? We probably need both: democracy (and design practices) are both a formal process of decision-making and problem-solving oriented towards a result, and a constant process of (local) negotiations and experimentation of new forms of social and political participation. Creating better ideas and 'solutions' and also new forms of common life, interactions, and participation.

Also, the two approaches have their strengths and weaknesses, which may determine the choice to invest in one or the other. For instance, Value-Sensitive Design and big public deliberation projects more naturally suit the approaches of big companies and governmental agencies and are more likely to attract their funding and support. Local experiments may be less likely to be supported by major political and economic institutions. This could be a consideration for academic designers and universities to support more local experiments. As DiSalvo puts it: 'during an era of "big data" – when everything must occur 'in the large' – the work of design and democracy can also be more modest; the context of the local and mundane is also important.'[84]

On the other hand, pragmatic experiments may indeed have a smaller impact and prove more difficult to scale up. They aim at being more inclusive than public deliberation projects by focusing on people and communities usually not involved in public deliberations. But in doing so, they may also not reach far beyond those communities, thereby arguably suffering from a different kind of lack of inclusivity.

However, it is not impossible, at least in principle, to look for projects and initiatives that contain some of the best advantages of both democratic theories and design approaches. As it were, creating a mixed model of designing for democracy combining more top-down coordination of large deliberation processes with local social experimentation; and, from the point of view of design, combining expert-led stakeholder involvement for the production of better systems with community-driven design experiments leading to new practices and institutions.

One promising example could be the original dialectic between governmental actors and civic tech activists in Taiwan. In 2014, the digital activist group g0v (pronounced gov-zero) played a critical role in coordinating the anti-government protests against a proposed trade deal with China, the so-called Sunflower Movement.[85] The protests led to a strong defeat of the governing Kuomintang party at successive municipal elections. The new Minister of Digital Affairs asked g0v volunteers to create an independent platform for online consultation: vTaiwan.[86] While vTaiwan in many respects is like e-Democracia and other digital deliberation platforms, the institutional arrangement created around it is not. In 2017, one of the most active promoters of the g0v movement, Audrey Tang, was appointed Minister for Digital Affairs. Since then, she has tried to bring her radical digital activist experience and mindset to the government of the country. Her most remarkable success was probably the management of the early stages of the COVID-19 pandemic. To address the spreading of the virus, she supported the independent project of some digital activists who had already been working on 'open data projects for citizens to interact with live maps, distributing ledge technology and chat bots to find the nearest pharmacy to claim free masks.'[87] This was arguably a successful attempt at creating a rapid, iterative, bottom-up process, where the government supports existing bottom-up civic experimental projects rather than devising its own top-down, government-led initiatives.

Notes

1 Evgeny Morozov, *The Net Delusion: The Dark Side of Internet Freedom* (New York: Public Affairs, 2011).
2 Pratyusha Kalluri, 'Don't Ask If Artificial Intelligence Is Good or Fair, Ask How It Shifts Power,' *Nature* 583, no. 7815 (1 July 2020): 169, https://doi.org/10.1038/D41586-020-02003-2.
3 'UN AI Actions,' *AI for Good* (blog), accessed 8 October 2023, https://aiforgood.itu.int/about-ai-for-good/un-ai-actions/.
4 'ACM FAccT Conference,' accessed 8 October 2023, https://facctconference.org/.
5 Shoshana Zuboff, *The Age of Surveillance Capitalism: The Fight for a Human Future at the New Frontier of Power* (London: Profile Books, 2019).
6 Morozov, *The Net Delusion.*
7 Thomas Metzinger, 'EU Guidelines: Ethics Washing Made in Europe,' *Tagesspiegel*, 4 August 2019, https://www.tagesspiegel.de/politik/eu-guidelines-ethics-washing-made-in-europe/24195496.html; Ben Wagner, 'Ethics As An Escape From Regulation. From "Ethics-Washing" To Ethics-Shopping?,' in *Being Profiled*, eds. Emre Bayamlioglu et al. (Amsterdam: Amsterdam University Press, 2019), 84–89, https://doi.org/10.1515/9789048550180-016.
8 See also Annette Zimmermann, Kate Vredenburgh, and Seth Lazar, 'The Political Philosophy of Data and AI,' *Canadian Journal of Philosophy* 52, no. 1 (2022): 1–5, https://doi.org/10.1017/can.2022.28; Mark Coeckelbergh, *The Political Philosophy of AI: An Introduction* (New York: Polity, 2022); Jonne Maas, 'A Neo-Republican Critique of AI Ethics,' *Journal of Responsible Technology* 9 (2022): 100022.
9 Michael Hardt and Antonio Negri, *Multitude: War and Democracy in the Age of Empire* (New York: Penguin Books, 2005).
10 Maša Galič, Tjerk Timan, and Bert-Jaap Koops, 'Bentham, Deleuze and Beyond: An Overview of Surveillance Theories from the Panopticon to Participation,' *Philosophy & Technology* 30, no. 1 (13 March 2017): 9–37, https://doi.org/10.1007/s13347-016-0219-1.

11 Michel Foucault, *Discipline and Punish: The Birth of the Prison* (New York: Vintage Books, 1995).
12 Galič, Timan, and Koops, 'Bentham, Deleuze and Beyond: An Overview of Surveillance Theories from the Panopticon to Participation.'
13 Gilles Deleuze, 'Postscript on the Societies of Control*,' *Surveillance, Crime and Social Control*, 1992, 35–39, https://doi.org/10.4324/9781315242002-3.
14 J. Matthew Hoye and Jeffrey Monaghan, 'Surveillance, Freedom and the Republic,' *European Journal of Political Theory* 17, no. 3 (2015): 343–63, https://doi.org/10.1177/1474885115608783.
15 For an overview of the debates in this area, see Coeckelbergh, *The Political Philosophy of AI: An Introduction.*
16 In this directions goes Jennifer Forestal, *Designing for Democracy : How to Build Community in Digital Environments* (Oxford: Oxford University Press, 2021).
17 Forestal.
18 C. Thi Nguyen, 'Echo Chambers and Epistemic Bubbles,' *Episteme* 17, no. 2 (June 2020): 141–61, https://doi.org/10.1017/epi.2018.32.
19 Eli Pariser, *The Filter Bubble: What the Internet Is Hiding from You* (London: Penguin Books, 2012).
20 More on this impact in chapter 12 below.
21 Kate Crawford, *Atlas of AI: Power, Politics, and the Planetary Costs of Artificial Intelligence* (New Haven: Yale University Press, 2021).
22 Julie Simon et al., *Digital Democracy: The Tools Transforming Political Engagement* (Nesta, 2017), 19, https://www.nesta.org.uk/report/digital-democracy-the-tools-transforming-political-engagement/.
23 Simon et al., 19.
24 Simon et al., 20.
25 Simon et al., 20.
26 Simon et al.
27 Carl DiSalvo, *Design as Democratic Inquiry: Putting Experimental Civics into Practice* (Cambridge, MA: The MIT Press, 2022), 103–38, https://doi.org/10.7551/MITPRESS/13372.001.0001.
28 DiSalvo, 103–38.
29 Batya Friedman, 'Value-Sensitive Design,' *Interactions* 3, no. 6 (December 1996): 16–23, https://doi.org/10.1145/242485.242493; Jeroen van den Hoven, Pieter E. Vermaas, and Ibo van de Poel, *Handbook of Ethics, Values, and Technological Design: Sources, Theory, Values and Application Domains* (Dordrecht: Springer, 2015), https://doi.org/10.1007/978-94-007-6970-0.
30 Peter M. Asaro, 'Transforming Society by Transforming Technology: The Science and Politics of Participatory Design,' *Accounting, Management and Information Technologies* 10, no. 4 (October 2000): 257–90, https://doi.org/10.1016/S0959-8022(00)00004-7.
31 Anthony Dunne and Fiona Raby, *Speculative Everything: Design, Fiction, and Social Dreaming* (Cambridge, MA and London: MIT Press, 2013); DiSalvo, 'Design as Democratic Inquiry.'
32 Jürgen Habermas, *The Structural Transformation of the Public Sphere: An Inquiry into a Category of Bourgeois Society*, 10. print, Studies in Contemporary German Social Thought (Cambridge, MA: The MIT Press, 1999).
33 John Rawls, 'The Idea of Public Reason Revisited,' *The University of Chicago Law Review* 64, no. 3 (1997): 765, https://doi.org/10.2307/1600311.
34 John Rawls, 'Two Concepts of Rules,' *The Philosophical Review* 64, no. 1 (1 January 1955): 3–32, https://doi.org/10.2307/2182230.

35 Thomas Scanlon, 'The Significance of Choice,' in *The Tanner Lectures on Human Values VIII* (Salt Lake City, UT: University of Utah Press, 1988), 178.
36 Amartya Sen, *Development as Freedom* (Oxford: Oxford University Press, 1999), 178.
37 Amartya Sen, *The Idea of Justice* (Cambridge, MA: Harvard University Press, 2009), 339–41.
38 Sen, 343–44.
39 Sen, *Development as Freedom*, 31.
40 Simon et al., *Digital Democracy: The Tools Transforming Political Engagement.*
41 This chapter focuses on the projects run by governmental and municipal agents, not political parties like Podemos in Spain, the Five Stat Movement in Italy and the Pirate Party in Germany and Scandinavian countries. These are also described and assessed in the Nesta report.
42 For a broader academic discussion of the connections between democratic theories and digital technologies, see Lucy Bernholz, Hélène Landemore, and Rob Reich, *Digital Technology and Democratic Theory* (Chicago: Chicago Press, 2021).
43 This point has been discussed in more general terms without specific reference to technology by Peter Mair, *Ruling the Void: The Hollowing of Western Democracy* (London and New York: Verso, 2013).
44 Simon et al., *Digital Democracy: The Tools Transforming Political Engagement.*
45 Simon et al., 46.
46 Simon et al., 44.
47 Simon et al., 30–31.
48 Simon et al., 67.
49 Simon et al., 67.
50 Simon et al., 83.
51 Trebor Scholz, 'Platform Cooperativism vs. the Sharing Economy,' *Medium .com*, 2014, 14, https://medium.com/@trebors/platform-cooperativism-vs-the-sharing-economy-2ea737f1b5ad.
52 Simon et al., *Digital Democracy: The Tools Transforming Political Engagement*, 92.
53 Jane Addams, *Twenty Years at Hull-House with Autobiographical Notes* (New York: The MacMillan Company, 1910), 110; DiSalvo, 'Design as Democratic Inquiry,' 25.
54 John Dewey, 'Challenge to Liberal Thought,' in *John Dewey: The Later Works, 1925-1953, Vol. 15, 1942–1948* (Carbondale, IL: Southern Illinois University Press, 1988), 273.
55 DiSalvo, 'Design as Democratic Inquiry,' 27.
56 Daniel Cefaï, 'Publics and Publicity: Towards a Pragmatist Enquiry,' *Politika*, 2016, https://www.politika.io/en/notice/publics-and-publicity-towards-a-pragmatist-enquiry.
57 Cefaï.
58 Cefaï.
59 DiSalvo, 'Design as Democratic Inquiry,' 15.
60 DiSalvo, 18.
61 DiSalvo, 16.
62 DiSalvo, 29–30.
63 DiSalvo, 136.
64 Jack Stilgoe, Richard Owen, and Phil Macnaghten, 'Developing a Framework for Responsible Innovation,' *Research Policy* 42, no. 9 (2013): 1568–80, https://doi.org/10.1016/j.respol.2013.05.008.
65 Jeroen van den Hoven, 'Value Sensitive Design and Responsible Innovation,' in *Responsible Innovation: Managing the Responsible Emergence of Science and*

Innovation in Society, eds. Richard Owen, John Bessant, and Maggy Heintz (Hoboken, NJ: John Wiley & Sons, Ltd, 2013), 75–83.

66 Friedman, 'Value-Sensitive Design.'

67 Rene von Schomberg, 'Towards Responsible Research and Innovation in the Information and Communication Technologies and Security Technologies Field,' *Luxemburg : Publication Office of the European Union, 2011* 20 March 2011, https://doi.org/10.2139/ssrn.2436399.

68 Jeroen van den Hoven, Gert-Jan Lokhorst, and Ibo Van de Poel, 'Engineering and the Problem of Moral Overload,' *Science and Engineering Ethics* 18, no. 1 (March 2012): 143–55, https://doi.org/10.1007/s11948-011-9277-z.

69 Van den Hoven, Lokhorst, and Van de Poel.

70 Asaro, 'Transforming Society by Transforming Technology.' See also Chapter 2 above.

71 Guy A. Boy, *The Handbook of Human-Machine Interaction: A Human-Centered Design Approach* (Boca Raton: CRC Press, 2017). See also Chapter 11 below.

72 Simon et al., *Digital Democracy: The Tools Transforming Political Engagement*, 20.

73 DiSalvo, 'Design as Democratic Inquiry,' 17.

74 Carl DiSalvo, *Adversarial Design* (Cambridge, MA: The MIT Press, 2012).

75 See also the approach of Transition Design: Terry Irwin, 'The Emerging Transition Design Approach,' *Design Research Society Conference* 2018, https://doi.org/10.21606/drs.2018.210.

76 Laura Forlano, 'Decentering the Human in the Design of Collaborative Cities,' *Design Issues* 32, no. 3 (1 July 2016): 42–54, https://doi.org/10.1162/DESI_a_00398.

77 Sasha Costanza-Chock, *Design Justice: Community-Led Practices to Build the Worlds We Need*, Information Policy (Cambridge, MA: The MIT Press, 2020).

78 Thomas Binder et al., 'Democratic Design Experiments: Between Parliament and Laboratory,' *CoDesign* 11, no. 3–4 (2 October 2015): 161, https://doi.org/10.1080/15710882.2015.1081248.

79 Binder et al., 162.

80 Binder et al., 161.

81 DiSalvo, 'Design as Democratic Inquiry,' 45.

82 DiSalvo, 160.

83 DiSalvo; Forlano, 'Decentering the Human in the Design of Collaborative Cities.'

84 DiSalvo, 'Design as Democratic Inquiry,' 18.

85 Simon et al., *Digital Democracy: The Tools Transforming Political Engagement*, 29.

86 Simon et al., 29.

87 Kelsie Nabben, 'Hacking the Pandemic: How Taiwan's Digital Democracy Holds COVID-19 at Bay,' *The Conversation*, 11 September 2020, http://theconversation.com/hacking-the-pandemic-how-taiwans-digital-democracy-holds-covid-19-at-bay-145023.

References

'ACM FACCT Conference.' Accessed 8 October 2023. https://facctconference.org/.

Addams, Jane. 'Twenty Years at Hull-House with Autobiographical Notes.' New York: The MacMillan Company, 1910.

AI for Good. 'UN AI Actions.' Accessed 8 October 2023. https://aiforgood.itu.int/about-ai-for-good/un-ai-actions/.

Asaro, Peter M. 'Transforming Society by Transforming Technology: The Science and Politics of Participatory Design.' *Accounting, Management and Information Technologies* 10, no. 4 (October 2000): 257–90. https://doi.org/10.1016/S0959-8022(00)00004-7.

Bernholz, Lucy, Hélène Landemore, and Rob Reich. *Digital Technology and Democratic Theory*. Chicago: Chicago Press, 2021.

Binder, Thomas, Eva Brandt, Pelle Ehn, and Joachim Halse. 'Democratic Design Experiments: Between Parliament and Laboratory.' *CoDesign* 11, no. 3–4 (2 October 2015): 152–65. https://doi.org/10.1080/15710882.2015.1081248.

Boy, Guy A. *The Handbook of Human-Machine Interaction: A Human-Centered Design Approach*. Boca Raton : CRC Press, 2017.

Cefaï, Daniel. 'Publics and Publicity: Towards a Pragmatist Enquiry.' *Politika*, 2016. https://www.politika.io/en/notice/publics-and-publicity-towards-a-pragmatist-enquiry.

Coeckelbergh, Mark. *The Political Philosophy of AI: An Introduction*. New York: Polity, 2022.

Costanza-Chock, Sasha. *Design Justice: Community-Led Practices to Build the Worlds We Need*. Cambridge, MA: The MIT Press, 2020.

Crawford, Kate. *Atlas of AI: Power, Politics, and the Planetary Costs of Artificial Intelligence*. New Haven: Yale University Press, 2021.

Deleuze, Gilles. 'Postscript on the Societies of Control*.' *Surveillance, Crime and Social Control* (1992): 35–39. https://doi.org/10.4324/9781315242002-3.

Dewey, John. 'Challenge to Liberal Thought.' In *John Dewey: The Later Works, 1925–1953, Vol. 15, 1942–1948*, 261–75. Carbondale, IL: Southern Illinois University Press, 1988.

DiSalvo, Carl. *Adversarial Design*. Cambridge, MA: The MIT Press, 2012.

———. *Design as Democratic Inquiry: Putting Experimental Civics into Practice*. Cambridge, MA: The MIT Press, 2022.

Dunne, Anthony, and Fiona Raby. *Speculative Everything: Design, Fiction, and Social Dreaming*. Cambridge, MA and London: MIT Press, 2013.

Forestal, Jennifer. *Designing for Democracy : How to Build Community in Digital Environments*. Oxford: Oxford University Press, 2021.

Forlano, Laura. 'Decentering the Human in the Design of Collaborative Cities.' *Design Issues* 32, no. 3 (1 July 2016): 42–54. https://doi.org/10.1162/DESI_a_00398.

Foucault, Michel. *Discipline and Punish: The Birth of the Prison*. New York: Vintage Books, 1995.

Friedman, Batya. 'Value-Sensitive Design.' *Interactions* 3, no. 6 (December 1996): 16–23. https://doi.org/10.1145/242485.242493.

Galič, Maša, Tjerk Timan, and Bert-Jaap Koops. 'Bentham, Deleuze and Beyond: An Overview of Surveillance Theories from the Panopticon to Participation.' *Philosophy & Technology* 30, no. 1 (13 March 2017): 9–37. https://doi.org/10.1007/s13347-016-0219-1.

Habermas, Jürgen. *The Structural Transformation of the Public Sphere: An Inquiry into a Category of Bourgeois Society*. Cambridge, MA: The MIT Press, 1999.

Hardt, Michael, and Antonio Negri. *Multitude: War and Democracy in the Age of Empire*. New York: Penguin Books, 2005.

Hoye, J. Matthew, and Jeffrey Monaghan. 'Surveillance, Freedom and the Republic.' *European Journal of Political Theory* 17, no. 3 (2018): 343–63. https://doi.org/10.1177/1474885115608783.

Irwin, Terry. 'The Emerging Transition Design Approach.' *Design Research Society Conference* 2018. https://doi.org/10.21606/drs.2018.210.

Kalluri, Pratyusha. 'Don't Ask If Artificial Intelligence Is Good or Fair, Ask How It Shifts Power.' *Nature* 583, no. 7815 (1 July 2020): 169. https://doi.org/10.1038/D41586-020-02003-2.

Maas, Jonne. 'A Neo-Republican Critique of AI Ethics.' *Journal of Responsible Technology* 9 (2022): 100022.

Mair, Peter. *Ruling the Void: The Hollowing of Western Democracy*. London and New York: Verso, 2013.

Metzinger, Thomas. 'EU Guidelines: Ethics Washing Made in Europe.' *Tagesspiegel*, 4 August 2019. https://www.tagesspiegel.de/politik/eu-guidelines-ethics-washing-made-in-europe/24195496.html.

Morozov, Evgeny. *The Net Delusion: The Dark Side of Internet Freedom*. New York: Public Affairs, 2011.

Nabben, Kelsie. 'Hacking the Pandemic: How Taiwan's Digital Democracy Holds COVID-19 at Bay.' *The Conversation*, 11 September 2020. http://theconversation.com/hacking-the-pandemic-how-taiwans-digital-democracy-holds-covid-19-at-bay-145023.

Nguyen, C. Thi. 'Echo Chambers and Epistemic Bubbles.' *Episteme* 17, no. 2 (June 2020): 141–61. https://doi.org/10.1017/epi.2018.32.

Pariser, Eli. *The Filter Bubble: What the Internet Is Hiding from You*. London: Penguin Books, 2012.

Rawls, John. 'The Idea of Public Reason Revisited.' *The University of Chicago Law Review* 64, no. 3 (1997): 765. https://doi.org/10.2307/1600311.

———. 'Two Concepts of Rules.' *The Philosophical Review* 64, no. 1 (1 January 1955): 3–32. https://doi.org/10.2307/2182230.

Scanlon, Thomas. 'The Significance of Choice.' In *The Tanner Lectures on Human Values*, Vol. VIII, 149–216. Salt Lake City, UT: University of Utah Press, 1988.

Scholz, Trebor. 'Platform Cooperativism vs. the Sharing Economy.' *Medium.com*, 2014. https://medium.com/@trebors/platform-cooperativism-vs-the-sharing-economy-2ea737f1b5ad.

Sen, Amartya. *Development as Freedom*. Oxford: Oxford University Press, 1999.

———. *The Idea of Justice*. Cambridge, MA: Harvard University Press, 2009. https://www.hup.harvard.edu/catalog.php?isbn=9780674060470.

Simon, Julie, Theo Bass, Victoria Boelman, and Geoff Mulgan. *Digital Democracy: The Tools Transforming Political Engagement*. Nesta, 2017. https://www.nesta.org.uk/report/digital-democracy-the-tools-transforming-political-engagement/.

Stilgoe, Jack, Richard Owen, and Phil Macnaghten. 'Developing a Framework for Responsible Innovation.' *Research Policy* 42, no. 9 (2013): 1568–80. https://doi.org/10.1016/j.respol.2013.05.008.

Van den Hoven, Jeroen. 'Value Sensitive Design and Responsible Innovation.' In *Responsible Innovation: Managing the Responsible Emergence of Science and Innovation in Society*, edited by Richard Owen, John Bessant, and Maggy Heintz, 75–83. Hoboken, NJ: John Wiley & Sons, Ltd, 2013.

Van den Hoven, Jeroen, Gert-Jan Lokhorst, and Ibo Van de Poel. 'Engineering and the Problem of Moral Overload.' *Science and Engineering Ethics* 18, no. 1 (March 2012): 143–55. https://doi.org/10.1007/s11948-011-9277-z.

Van den Hoven, Jeroen, Pieter E. Vermaas, and Ibo van de Poel. *Handbook of Ethics, Values, and Technological Design: Sources, Theory, Values and Application Domains*. Dordrecht: Springer, 2015. https://doi.org/10.1007/978-94-007-6970-0.

Von Schomberg, René. 'Towards Responsible Research and Innovation in the Information and Communication Technologies and Security Technologies Field.' Luxemburg : Publication Office of the European Union, 2011, 20 March 2011. https://doi.org/10.2139/ssrn.2436399.

Wagner, Ben. 'Ethics as an Escape from Regulation: From "Ethics-Washing" to Ethics-Shopping?' In *Being Profiled*, edited by Emre Bayamlioglu, Irina Baraliuc, Liisa Albertha Wilhelmina Janssens, and Mireille Hildebrandt, 84–89. Amsterdam University Press, 2019. https://doi.org/10.1515/9789048550180-016.

Zimmermann, Annette, Kate Vredenburgh, and Seth Lazar. 'The Political Philosophy of Data and AI.' *Canadian Journal of Philosophy* 52, no. 1 (2022): 1–5. https://doi.org/10.1017/can.2022.28.

Zuboff, Shoshana. *The Age of Surveillance Capitalism: The Fight for a Human Future at the New Frontier of Power*. London: Profile Books, 2019.

11 Expanding democracy

Design for aesthetic engagement and anti-power

11.1 Beyond argumentation and representation

Michael Christopher Brown, a US photojournalist famous for documenting life and conflicts in Africa and the Middle East, has recently used the artificial intelligence (henceforth AI) image generator Midjourney to produce a series of images – at first glance hardly distinguishable from real photographs – representing episodes of migrations of Cuban people towards Florida.[1] He wanted to make visible to the public a phenomenon of which there are no images: people fleeing from the Cuban regime.

In the ongoing Syrian war, 'there are more hours of videos documenting the conflict than there have been hours in the conflict itself. In 2019, eight years after the Syrian conflict began in 2011, more than 50 videos were uploaded to YouTube each day.'[2] The Syrian Archive has organised this immense amount of digital material into a well-ordered archive, one that might be used among other things for holding the Syrian regime accountable for their war crimes against its population and their human rights violation.

These two stories briefly illustrate two possible further directions for 'designing for democracy.' First, Brown's project suggests expanding the idea of the 'public sphere' to explicitly include also 'non-argumentative' contributions, namely aesthetic, artistic, and expressive forms of engagement, and to support the design of projects of this kind. I will call this approach 'more-than-argumentative' democracy; the Syrian Archive suggests a way of expanding democratic participation from deliberation to 'anti-corruption' and 'power control' and designing systems that may support counter-powers, even in non-democratic, authoritarian political regimes. I will call this 'more-than-representative' democracy.

Theories of deliberative or participatory democracy presented so far do not seem to sufficiently cover these different ways of representation and power control. Thankfully, paraphrasing Shakespeare, there are more theories of democracy than one single philosophy may come up with. If one wants to design for democratic AI, it is therefore important to be

DOI: 10.4324/9781003303244-15

aware of the many – sometimes complementary, sometimes conflicting – views of what democracy is or should be. I will look both at theories of democracy in traditional political philosophy as well as the contribution of design theories. As usual, for each theoretical approach, I will present examples of possible applications. While this chapter focuses on 'more-than-argumentative' and 'more-than-representative' democracy, the next chapter will propose two more directions: post-colonial and more-than-human democracy.

11.2 More-than-argumentative democracy

Deliberative democracy theory grounded in rational deliberation as presented above and exemplified by open government projects like e-Democracia has received one major critique: it disproportionately favours rational argumentation as the main or sole legitimate communicative means of participation in deliberation.[3] Rational argumentation 'posits a reflexive, impartial, reasoned exchange of validity claims where only the force of better argument "wins out."'[4] It is typically promoted in (Western) rationalistic philosophies and realised, for instance, in legal adjudication. It usually relies on the well-ordered presentation of arguments in support of a certain thesis, the discussion and disproof of possible counter-arguments, the connection of a claim to a more general principle that is accepted as valid, etc. An exclusive focus on this style of deliberation as constitutive of the public sphere is problematic for two connected reasons: first, it tends to exclude people and groups who are not able or willing to express themselves and communicate in that way; second, and relatedly, it tends to impoverish the public debate and the deliberation process by narrowing down its possible forms. Contributions that are typically marginalised in an ideal rational-deliberative process are those based on an emotional-affective commitment and those that are expressed in less argumentative terms, for instance, through artistic means, parody, storytelling, or testimonials.

It is an open question to what extent deliberative democracy theory and practice can be expanded beyond its narrow, argumentative commitments to include these alternative forms of contributions. This is what supporters of the so-called 'difference democracy' like Seyla Benhabib and Iris Young claim and plea for.[5] According to 'difference democrats,' democracy must become more inclusive and should create the possibility for non-argumentative, namely emotional-affective and aesthetic contributions to be heard and to matter more in public deliberation.

However, other political philosophers claim that deliberative democracy theory cannot sufficiently include these non-argumentative forms of deliberation. From their perspective, non-argumentative contributions must be deployed not as part of the public debate but rather to shake its foundations, to show its limitations, and to highlight the (unjust) structures of powers that underline it. These are the supporters of the so-called 'adversarial' or 'agonistic' democracy.[6] In their view, argumentative discourse is not only a means of communication but rather the manifestation of a structure of power – typically, in Western countries, of male, white, wealthy, and educated groups – which must be unmasked, challenged, and replaced rather than just reformed, made more inclusive or enriched. Often influenced by feminism and Foucault's reflection on disciplinary power, adversarial democrats hold that accepting any of the 'rules of communication' typically associated with rational democratic deliberation means supporting the power of the groups who have traditionally defined and benefitted from those rules. In the Foucauldian words of Dana Villa: 'to act according to a norm is to be normalized in some sense.'[7]

I cannot address – let alone solve – this philosophical and political debate here. For the sake of this chapter, I will just rely on the general idea shared by these traditions: that deliberative democracy narrowly conceived – happening in the public sphere through an orderly, rational exchange of arguments – is not sufficient to give power and representation to all groups in society and to create a sufficiently rich and meaningful public discourse. There must be more space for aesthetic engagement and forms of communication grounded in intuition, imagination, and artistic production.

To be sure, such forms of expression already exist and are very prominent in people's everyday lives and in the media. However, they are often not sufficiently recognised – in political theory and practice – as legitimate forms of contributions to the political debate or to the deliberation process over public issues. Of course, there are good reasons to be cautious with the expansion of democratic deliberation towards including emotional-affective engagement. Right-wing populism, conspiracy theories, fake news, and hate speech are very much fed by this kind of engagement. So, one may rather be tempted to reinforce rational deliberation than to expand emotional and affective engagement. However, I rather propose that we should try and design for a better emotional and aesthetic engagement.

Referring to Hannah Arendt's late writings on the issue of judgement, pragmatist philosopher Daniel Cefaï emphasises the importance of aesthetic and affective experience – as opposed to the mere rational argumentation – for the creation of 'publics,' that is, groups of citizens following and engaged in specific political issues, able to have a meaningful contribution to political life and deliberation:

> The collective activity of publics requires the strength of calling forth testimonials, the visionary power of utopia, the sharing of feelings of justice, and the contagion of the trials of revolt. … these affective perceptive and moral trials can be found just as much in the uproar

> of concertation meeting whose audience revolts, as in the handling of a case before a labour tribunal, or in the stories of witnesses recorded by a documentary filmmaker. Publicity proliferates, is disseminated and propagated in all these situations.[8]

The normative force and legitimacy of public reason derives not only from the rational force of the arguments proposed but also from the 'expressive legitimacy of collective demands.'[9] Contributions based on imagination, empathy, and emotion rather than rational argumentation – think of film-making, songs, novels, and theatre – can also be valid, to the extent that they express some legitimate needs, aspirations, or demands of a group in society.[10]

As I will show later while discussing the importance of mechanisms of 'anti-power' in democracies, artistic forms of political engagement may be particularly effective and desirable to contrast the power of authoritarian regimes that do not allow for inclusive parliamentary deliberations, free press, or other forms of argumentative contributions. One famous example is the 'Theatre of the Oppressed' genre, created in Brazil in the 1970s by the playwright Augusto Boal,[11] inspired by the book *Pedagogy of the Oppressed* by Paulo Freire.[12] It involves interactive theatre performances presenting scenarios of social oppressions where audiences are invited to intervene and practice. The goal is to prompt people to imagine and enact forms of social resistance and political action, by rehearsing and practising in the relatively safe space of a theatrical performance. A variation elaborated by Boal during his stay in Argentina in the 1970s is the 'Invisible Theatre,' where the performances happen in disguise in the street or other public spaces. Actors do not disclose that they are performing, and engaged passers-by are not told that they are part of a theatre performance.[13] The Invisible Theatre lends itself well to be enacted in countries where official forms of protests and possibly even politically engaged performances are not allowed or subject to governmental censorship.

For sure, the positive power of affective, aesthetic, and artistic expressions in political discussions has been increasingly recognised within formal, institutional settings. Consider, as one example, the issue of migrations in Italy. Since 2001, almost 30,000 people have died while attempting to cross the Mediterranean to reach Europe from Africa. Many of the victims could not be identified. In 2018, the medical examiner Cristina Cattaneo published a book in which she tried to give back an identity to some of the persons who died at sea, after creating a database to collect information about them: things found on their bodies and information coming from their families. Among the many tragic and touching stories she collected was that of one young boy of around 14 years old from Mali, found dead following one of the shipwrecks. He was travelling without documents but with a school report in Arabic and French sewn into his pocket. He might have wanted to bring this as a proof of his goodwill to come to Europe to study. The graphic artist Maccox captures this unbearably tragic story in a cartoon, published on the newspaper *Il Foglio*, next to an article presenting Cattaneo's book.[14] The cartoon shows a curled-hair boy, sitting alive on the seabed, proudly presenting his school report to a colourful marine wildlife. Looking at the report, a big purple octopus comments: 'Wow ... all tens... a rare pearl.' The cartoon is titled *Lost Treasures*. Recently, Italian President of the Republic Sergio Mattarella said in a public speech that that cartoon, more than any fact and figure, helped him better understand migration, by vividly reminding him of the 'uncountable, individual people, each with their stories, their plans, their dreams, their future.'[15]

As already seen in the previous chapter, broadening the scope of what should count as a democratic deliberation or democratic participation beyond the mere exchange of rational arguments is important not only theoretically but also from the practical perspective of 'design for democracy.' By expanding the repertoire of legitimate means of participation in deliberative democracy we can expand the repertoire of designed tools and processes to support and improve this engagement.

11.3 AI, new media, and civil engagement

Artists have always been craftspeople in the first place. Painters, sculptors, musicians, photographers, and film directors have used technical tools to produce aesthetically valuable experiences through the production of objects. Artworks have sometimes also had the explicit goal of making political statements and exposing political injustices. Think of Picasso's painting *Guernica* as a visual statement about the barbary of the bombing of the Spanish city of the same name done by Fascist and Nazi aviation in support of Franco in the Spanish civil war; or Stanley Kubrick's *A Full Metal Jacket*, an unsettling and grotesque film about the psychological impact of war traumas on soldiers. Other genres, like science fiction, have sometimes functioned more as a trigger for an open societal reflection and discussion on the relationship between humanity and technology. Think of the recent Netflix show *Black Mirror*, a series of episodes set in a not-so-distant future where current technologies such as AI, robotics, and neuroimplants have been further developed and are everywhere in people's everyday life.

By looking at artists as designers, we could say that the first approach – art as a political statement – is somehow close to Value-Sensitive Design, where specific values and principles are intentionally designed into an object, while the second – art and science fiction as a reflection prompt – would be closer to Critical or Speculative Design, where objects are used to spark a more open societal and political debate about the future of humanity and technology.

The importance of affective and emotional contributions to the deliberation about technology has also been recognised in philosophy of technology. Sabine Roeser calls for a re-evaluation of moral emotions like fear, indignation and compassion in public debates over risky technologies.[16] Starting from the empirically supported consideration that moral emotions may support valid moral judgements and valuable political arguments, she makes a plea for an 'emotional deliberation approach' to risky technologies. Art and aesthetic experiences can play a big role in eliciting these emotions, therefore enriching the debates on the future of technology.[17]

Another important role that art and aesthetic experience can play is giving visibility and voice to people who cannot otherwise participate and speak in the political debate like, for instance, migrants. Technological development allows for a broader variety of projects. In the Vía PanAm project, photographer Kadir van Lohuizen documented his 40-week journey from Southern Chile to Northern Alaska, telling the stories of migrants he met along the way, through a mix of old and new media: blogs, photo reports, video, audio, and apps.[18]

In a more controversial project, Michael Christopher Brown, a photojournalist famous for his reportages about life and conflicts in Africa and the Middle East, used the AI image generator Midjourney to produce a series of images – at a first superficial glance hardly distinguishable from real photographs – representing episodes of migration of Cuban people fleeing from the Cuban regime towards Florida.[19] Many found the project unethical as it arguably risked to dangerously blur the line between reality and fiction in politically sensitive reportages and documents. Also, Brown was criticised for selling the 400 images for profit. He stood by his project, saying it was a way to make visible true stories that he cannot access with his camera. And that he has always labelled the images as 'not real' on his Instagram posts.[20] Also, Brown said that he was pleased about the heated ethical debates and criticism raised by his project, as it aims 'to surface public conversation and questioning around what work of this nature may mean for image-based storytellers who care about reality and truth' in a time when, thanks to AI advancements, 'the line between what is real and what is fake could become increasingly opaque.'[21]

Brown's story highlights the opportunities and risks of using new media and AI for political purposes. While his intention was to document a true story, new media are often used to spread misinformation – fake news – and support political agendas of authoritarian political regimes: 'cognitive warfare.'[22]

A first design challenge is supporting the political use of new (AI) media in ways that promote, not endanger, an open and inclusive democratic debate and deliberation.

Brown's project is also very close to an experiment in Critical or Speculative design, and this is a second direction to follow: promoting projects that raise concerns, spark debate and support the imagination of possible (desirable) futures, and the role of new media in promoting or weakening democratic debates.

11.4 More-than-representative democracy

I have previously introduced the idea of democracy as inclusive representation. However, representation is only one defining character of democracy. Historically, democratic procedures and institutions have arguably had at least two main goals: to represent people and to control governmental power, that is, 'anti-corruption'[23] or 'anti-power.'[24] Broadly conceived, corruption is the use of public or State power to benefit partial or private interests. It is an abuse of public power. While there is a large overlap between the demands of representation and anti-corruption, these are not fully interchangeable. On the one hand, representativeness is insufficient to prevent all forms of corruption. A good representation of different groups in a deliberation room may prevent many forms of self-serving decisions by specific groups at the expenses of others. However, representation cannot eliminate all forms of corruption. No one is immune to the lure of abusing public power for one own's interest and corruption is likely to persist even when institutions are representative: people who are elected to serve the interests of a group they belong to may also turn to self-serving behaviour. On the other hand, representativeness is also *not necessary* for anti-corruption, that is, anti-corruption can also be achieved through institutions that are not otherwise representative of the population. Both Athenians and Medieval Italian republics originally introduced lotteries to choose public officials to break up concentrations of power. The idea was that, unlike powerful elected officials, randomly chosen persons would be less likely to so-called 'power capture,' that is, to express the partisan interests of the elites they represented. Similarly, lotteries were introduced for jury trials in England (and its American colonies) to insulate judicial processes against interference by the powerful.[25] In more recent times, we see civic associations, watchdogs, and independent journalists – all actors that do not necessarily aim to represent the people – to perform an indispensable role of anti-corruption in democracies as well as in countries with non-democratic regimes.

Political philosopher Uğur Aytaç remarks that democratic theories – including the ones discussed above: deliberative democracy in its different forms and neo-republican democracy – tend to disagree on issues like the definition of the public sphere, what counts as inclusive deliberation, representation, and so forth; but they tend to share the demand that the public sphere should be autonomous from the concentrated powers of social, economic, and political elites.[26]

He suggests that the idea of anti-corruption is helpful to understand and address the issue of the growing political power of privately owned internet platforms in shaping the public sphere. As seen above, privately owned digital platforms are key spaces for public deliberation and political engagement. Big Tech's platforms have become gatekeepers in the new media industry: in 2017, 70% of news websites' external traffic was on Google and Facebook. These platforms are not neutral, open spaces where people spontaneously gather to read and discuss, as in the Hayekian depiction of market relations.[27] They are rather 'built environment,' consciously designed to promote certain kinds of interactions.[28] As discussed in Chapter 5, Big Tech's business crucially depends on getting people interacting as much as possible on their platform, a goal they achieve through personalisation of content and by inducing emotional attachment in the users. While the precise effect of this algorithmic curation on civic engagement – illustrated by phenomena like 'filter bubbles'[29] and 'echo chambers'[30] – is still open to study and debate, some facts are relatively uncontroversial, such as that personalisation through algorithmic curation tends to promote mobilisation around high-profile users such as Donald Trump. This affects the capacity of citizens as well as journalists to produce and engage with publicly relevant content. They are powers that affect the entire democratic life.[31]

Digital platform and the algorithms on which they run have thus become a key institutional power shaping the public space. However, seen as public institutions, they are, as it were, substantially 'corrupted.' In their current form, far from being designed to provide a space that is conducive to a meaningful and inclusive public deliberation, they are rather designed to support a strong concentration of attention and power by some elites in the public sphere.

How this specific new form of corruption of the public sphere can be limited and controlled is an open and thorny question. There seem to be three main directions:[32] first, simply breaking the power of the private platform by making them smaller, a solution that would certainly address the problem of the concentration of power but would also sacrifice the existence of a big shared digital space for information and engagement, the possibility, as it were, of a global public sphere; second, implementing strong forms of State regulation about the functioning of the private platforms, another potentially effective solution that is, however, open to forms of power capture by the government and the political parties that hold its power; third, in the spirit of (non-representative) anti-corruption mechanisms discussed in this chapter, having small, independent, boards of citizens appointed to assess, control, and reform the functioning of the platform algorithms.[33] While not being representative, these boards may certainly be effective, to the extent that their relatively few members could receive a specific training and be committed to only this one task for a given period. They would not be professional, elected politicians with personal longer-term agendas and dependent on the support of the electorate and the funds of private actors.

While technical systems and 'objects' acquire more and more explicit and pervasive ethical and political power, we should wonder how we can make them track the reasons and interests of the relevant human agents they are meant to serve. This is what I have discussed under the idea of 'Meaningful Human Control' over AI in the previous chapters. This is also, roughly, the idea behind some theories of liberal democracy: public power should track or represent the interests of people. In this case, a public power embedded in technological systems. The focus on anti-corruption allows one to see a complementary strategy: creating political mechanisms to break the power's tendency to be corrupted, that is to be captured and directed toward the interests of some powerful elites; and devising appropriate institutional design to counteract the corruption of socio-technical systems.

The idea of anti-corruption supports yet another, complementary strategy: designing socio-technical systems that can support control and opposition to corrupted political governments.

11.5 Anti-power technologies

In *The Net Delusion* (2011), the journalist and activist Evgeni Morozov criticised the naiveté of what he called 'cyber-utopianism,' the idea that new media would free up and democratise the world, as it were, automatically, only by spreading information. The early 2000s had seen the rise and partial fall of the promise of the internet mainstream platforms promoting effective and lasting anti-regime revolutions, the so-called Arab Spring.[34] For a moment, many believed that revolutions would be 'twittered.' Morozov himself admitted to having been a victim of the lure of cyber-utopianism, when he joined as a young Belarus man a Western NGO that wanted to promote democracy in the former Soviet bloc with the help of the internet. But after a few years he noticed that in non-democratic countries the internet was not effective in challenging the government's power. To the contrary, governments had started to learn how to control the internet by censoring the blogs and the social networks of the activists and even to actively use the new media themselves to spread governmental propaganda and to collect information about their opponents. A story that speaks to the present, looking at the risks for politically engaged journalists and citizens of non-democratic countries,[35] and the massive use of digital media made by authoritarian regimes to control opponents and gain support ('cognitive warfare').[36]

Off-the-shelf technology does not spontaneously create freedom and democracy, and if not properly designed can even backfire. Back in 2011,

Morozov wondered, 'How hard is to imagine a site like Facebook inadvertently disclosing the private information of activists in Iran or China, tipping off the governments to secret connections between the activists and their Western funders?'[37]

To be effective as a tool against non-democratic or democratic but corrupted powers, technology must be designed to respond to the specific local needs of the political action. One among many recent examples in this direction is the Syrian Archive.

> The Syrian Archive aims to support human rights investigators, advocates, media reporters, and journalists in their efforts to document human rights violations in Syria and worldwide through developing new open source tools as well as providing a transparent and replicable methodology for collecting, preserving, verifying and investigating visual documentation in conflict areas. ... the Syrian Archive aims to preserve data as a digital memory, to establish a verified database of human rights violations, and to act as an evidence tool for legally implementing justice and accountability as concept and practice in Syria.[38]

The project has many objectives such as to 'humanise victims, reduce the space for dispute over numbers killed, help societies understand the true human costs of war, and support truth and reconciliation efforts.'[39] Among these goals is also collecting evidence for holding powers accountable for their human rights violation, as it were, creating a space for democratic resistance in a conflict largely controlled by a ruthless authoritarian regime and its international supporters.

Such evidence can be in the form of visual documentation. For centuries, professional journalists and photographers have tried to capture and document with their cameras the horrors of war and the systematic violations of human rights. In 1972, US Associated Press photographer Nick Ut captured the horrifying scene of a group of children fleeing a Napalm attack from the US army on the village of Trang Bang, in Vietnam. At the centre of his picture, a nine-year-old girl runs and screams in pain and terror, fully naked. Phan Thi Kim Phuc, the girl in the picture, went on to eventually survive her injuries, thanks to the help of the photographer who rushed to her and others' aid immediately after capturing their distress on camera. Years later, she explained that she had stripped her clothes off after they were caught on fire by the bombing, like everything else in the village. And that she was screaming in Vietnamese: 'Too hot! Too hot!' That picture has remained for many the 'defining photo' of the Vietnam War and a warning about the war's horror. Yet, it did not have a big impact on the conflict itself, and there is no evidence that it accelerated its end or had any significant impact on the already critical of war US public opinion at the time of its publication.[40]

Today, in the Syrian conflict, 'there are more hours of videos documenting the conflict than there have been hours in the conflict itself. In 2019, eight years after the Syrian conflict began in 2011, more than 50 videos were uploaded to YouTube each day.'[41] This immense 'accidental archive,' made possible by the existence of widely available technological tools and platforms like smartphones and social media platforms, has a big ethical, legal and political potential. This potential must be unlocked by a professional and piecemeal curation of more specific, context-specific, tools and techniques. In a sort of Value-Sensitive Design process, different values, sometimes in potential tension, must be accounted for: the public availability of information, the recognition of the moral demands of victims, the protection of the authors of the videos from the revenge of the government, and the future usability of the materials in formal procedures like International Tribunals and others.

The value of anti-corruption and anti-power technologies looks particularly evident in countries where other forms of democratic participation are difficult or impossible. However, they can, must, and in fact do play an important role in more democratic regimes. As highlighted in the neo-republican and other political traditions, the role of civic associations, watchdogs, and independent journalists in power control is crucial for keeping democracies healthy. To mention one example among many possible others, think of the work of the recently assassinated journalist and blogger Daphne Caruana Galizia, famous for her work of investigation and exposure of the Maltese government's corruption.[42] Researchers should study what the role of new technologies can be in supporting and promoting, possibly protecting the work of these anti-power agents also in democratic countries.

Notes

1 '90 Miles,' accessed 4 October 2023, https://www.airlab.co/90-miles.
2 'Syrian Archive,' accessed 4 October 2023, https://syrianarchive.org/.
3 Lincoln Dahlberg, 'The Habermasian Public Sphere: Taking Difference Seriously?,' *Theory and Society* 34, no. 2 (1 April 2005): 111–36, https://doi.org/10.1007/s11186-005-0155-z.
4 Dahlberg, 113.
5 Seyla Benhabib, 'Toward a Deliberative Model of Democratic Legitimacy,' in *Democracy and Difference: Contesting the Boundaries of the Political*, ed. Seyla Benhabib (Princeton, NJ: Princeton University Press, 1996); Iris Marion Young, *Inclusion and Democracy* (Oxford: Oxford university press, 2000).
6 Chantal Mouffe, *The Democratic Paradox*, Repr (London and New York: Verso, 2009); Carl DiSalvo, *Adversarial Design* (Cambridge, MA: The MIT Press, 2012).
7 Dana R. Villa, 'Postmodernism and the Public Sphere,' *American Political Science Review* 86, no. 3 (September 1992): 715, https://doi.org/10.2307/1964133; Dahlberg, 'The Habermasian Public Sphere,' 122.

8 Daniel Cefaï, 'Publics and Publicity: Towards a Pragmatist Enquiry,' *Politika*, 2016, https://www.politika.io/en/notice/publics-and-publicity-towards-a-pragmatist-enquiry.
9 Cefaï.
10 Cefaï.
11 Augusto Boal, *Theater of the Oppressed* (New York: Theater Communication Group, 1993).
12 Paulo Freire, *Pedagogy of the Oppressed*, 30th anniversary ed. (New York: Continuum, 2000).
13 Augusto Boal, 'Invisible Theater,' in *Radical Street Performance: An International Anthology*, ed. Jan Cohen-Cruz (London and New York: Routledge, 1994), 121–24.
14 Nicola Scevola, 'Senza nome in mare,' *Il Foglio*, 1 November 2019, https://www.ilfoglio.it/il-figlio/2019/01/11/news/senza-nome-in-mare-232521/.
15 Marina Corradi, 'Mattarella e il disegno del ragazzino morto in mare. Voleva studiare in Europa,' *Avvenire*, 26 August 2023, https://www.avvenire.it/attualita/pagine/mattarella-disegno-bambino-migrante-in-fondo-al-mare-con-la-pagella.
16 Sabine Roeser, *Risk, Technology, and Moral Emotions* (New York and London: Routledge, 2018).
17 Sabine Roeser, Veronica Alfano, and Caroline Nevejan, 'The Role of Art in Emotional-Moral Reflection on Risky and Controversial Technologies: The Case of BNCI,' *Ethical Theory and Moral Practice* 21, no. 2 (1 April 2018): 275–89, https://doi.org/10.1007/s10677-018-9878-6.
18 Kadir Van Lohuizen, 'Vía PanAm: A 40-week Journey Exploring Migration in the Americas,' (Digital Publication) Ydoc Pub. https://www.paradox.nl/project/via-panam/#work 2011, 420.
19 '90 Miles.'
20 Matt Growcoot, 'Photojournalist Controversially Turns to AI to Illustrate "Inaccessible" Stories,' *PetaPixel*, 15 May 2023, https://petapixel.com/2023/05/15/photojournalist-controversially-turns-to-ai-to-illustrate-inaccessible-stories/.
21 Growcoot.
22 Seumas Miller, 'Cognitive Warfare: An Ethical Analysis,' *Ethics and Information Technology* 25, no. 3 (4 September 2023): 46, https://doi.org/10.1007/s10676-023-09717-7.
23 Samuel Bagg, 'Sortition as Anti-Corruption: Popular Oversight against Elite Capture,' *American Journal of Political Science*, 2022, https://doi.org/10.1111/AJPS.12704.
24 Philip Pettit, 'Freedom as Antipower,' *Ethics* 106, no. 3 (April 1996): 576–604, https://doi.org/10.1086/233648.
25 Bagg, 'Sortition as Anti-Corruption: Popular Oversight against Elite Capture.'
26 Ugur Aytaç, 'Big Tech, Algorithmic Power, and Democratic Control,' *The Journal of Poitics* (forthcoming).
27 Aytaç.
28 Jennifer Forestal, *Designing for Democracy: How to Build Community in Digital Environments* (Oxford: Oxford University Press, 2021).
29 Eli Pariser, *The Filter Bubble: What the Internet Is Hiding from You* (London: Penguin books, 2012).
30 C. Thi Nguyen, 'Echo Chambers and Epistemic Bubbles,' *Episteme* 17, no. 2 (June 2020): 141–61, https://doi.org/10.1017/epi.2018.32.
31 Aytaç, 'Big Tech, Algorithmic Power, and Democratic Control.'
32 Aytaç.

33 Aytaç.
34 'What Is the Arab Spring, and How Did It Start?,' *Al Jazeera*, accessed 3 October 2023, https://www.aljazeera.com/news/2020/12/17/what-is-the-arab-spring-and-how-did-it-start.
35 'Jamal Khashoggi: All You Need to Know about Saudi Journalist's Death,' *BBC News*, 10 October 2018, sec. Europe, https://www.bbc.com/news/world-europe-45812399.
36 Miller, 'Cognitive Warfare.'
37 Evgeny Morozov, *The Net Delusion: The Dark Side of Internet Freedom* (New York: PublicAffairs, 2011), xv.
38 'Syrian Archive.'
39 'Syrian Archive.'
40 Oscar Holland, '"Napalm Girl" at 50: The Story of the Vietnam War's Defining Photo,' 6 September 2022, https://edition.cnn.com/style/article/napalm-girl-50-snap/index.html#.
41 'Syrian Archive.'
42 Joseph A. Debono, ed., *Invicta: The Life and Work of Daphne Caruana Galizia* (Oxford: Pertinent Press, 2017).

References

'90 Miles.' Accessed 4 October 2023. https://www.airlab.co/90-miles.

Aytaç, Ugur. 'Big Tech, Algorithmic Power, and Democratic Control.' *The Journal of Politics*, forthcoming.

Bagg, Samuel. 'Sortition as Anti-Corruption: Popular Oversight against Elite Capture.' *American Journal of Political Science*, 2022. https://doi.org/10.1111/AJPS.12704.

BBC News. 'Jamal Khashoggi: All You Need to Know about Saudi Journalist's Death.' 10 October 2018, sec. Europe. https://www.bbc.com/news/world-europe-45812399.

Benhabib, Seyla. 'Toward a Deliberative Model of Democratic Legitimacy.' In *Democracy and Difference: Contesting the Boundaries of the Political*, edited by Seyla Benhabib. Princeton, NJ: Princeton University Press, 1996.

Boal, Augusto. 'Invisible Theater.' In *Radical Street Performance: An International Anthology*, edited by Jan Cohen-Cruz, 121–24. London and New York:Routledge, 1994.

———. *Theater of the Oppressed*. New York : Theater Communication Group, 1993.

Cefaï, Daniel. 'Publics and Publicity: Towards a Pragmatist Enquiry.' *Politika*, 2016. https://www.politika.io/en/notice/publics-and-publicity-towards-a-pragmatist-enquiry.

Corradi, Marina. 'Mattarella e il disegno del ragazzino morto in mare. Voleva studiare in Europa.' *Avvenire*, 26 August 2023. https://www.avvenire.it/attualita/pagine/mattarella-disegno-bambino-migrante-in-fondo-al-mare-con-la-pagella.

Dahlberg, Lincoln. 'The Habermasian Public Sphere: Taking Difference Seriously?' *Theory and Society* 34, no. 2 (1 April 2005): 111–36. https://doi.org/10.1007/s11186-005-0155-z.

Debono, Joseph A., ed. *Invicta: The Life and Work of Daphne Caruana Galizia*. Oxford: Pertinent Press, 2017.
DiSalvo, Carl. *Adversarial Design*, Cambridge, MA: The MIT Press, 2012.
Forestal, Jennifer. *Designing for Democracy: How to Build Community in Digital Environments*. Oxford: Oxford University Press, 2021.
Freire, Paulo. *Pedagogy of the Oppressed*, 30th anniversary ed. New York: Continuum, 2000.
Growcoot, Matt. 'Photojournalist Controversially Turns to AI to Illustrate 'Inaccessible' Stories.' *PetaPixel*, 15 May 2023. https://petapixel.com/2023/05/15/photojournalist-controversially-turns-to-ai-to-illustrate-inaccessible-stories/.
Holland, Oscar. '"Napalm Girl" at 50: The Story of the Vietnam War's Defining Photo.' *CNN*, 6 September 2022. https://edition.cnn.com/style/article/napalm-girl-50-snap/index.html#.
Miller, Seumas. 'Cognitive Warfare: An Ethical Analysis.' *Ethics and Information Technology* 25, no. 3 (4 September 2023): 46. https://doi.org/10.1007/s10676-023-09717-7.
Morozov, Evgeny. *The Net Delusion: The Dark Side of Internet Freedom*. New York: PublicAffairs, 2011.
Mouffe, Chantal. *The Democratic Paradox*. Repr. Radical Thinkers. London and New York: Verso, 2009.
Nguyen, C. Thi. 'Echo Chambers and Epistemic Bubbles.' *Episteme* 17, no. 2 (June 2020): 141–61. https://doi.org/10.1017/epi.2018.32.
Pariser, Eli. *The Filter Bubble: What the Internet Is Hiding from You*. London: Penguin Books, 2012.
Pettit, Philip. 'Freedom as Antipower.' *Ethics* 106, no. 3 (April 1996): 576–604. https://doi.org/10.1086/233648.
Roeser, Sabine. *Risk, Technology, and Moral Emotions*. Routledge Studies in Ethics and Moral Theory 39. New York and London: Routledge, 2018.
Roeser, Sabine, Veronica Alfano, and Caroline Nevejan. 'The Role of Art in Emotional-Moral Reflection on Risky and Controversial Technologies: The Case of BNCI.' *Ethical Theory and Moral Practice* 21, no. 2 (1 April 2018): 275–89. https://doi.org/10.1007/s10677-018-9878-6.
Scevola, Nicola. 'Senza Nome in Mare.' *Il Foglio*, 1 November 2019. https://www.ilfoglio.it/il-figlio/2019/01/11/news/senza-nome-in-mare-232521/.
'Syrian Archive.' Accessed 4 October 2023. https://syrianarchive.org/.
Van Lohuizen, Kadir . 'V.a PanAm: Exploring Migration in the Americas = Explorando Las Migraciones En Las Am.ricas.' (Digital Publication) Ydoc Pub. https://www.paradox.nl/project/via-panam/#work (2011). http://www.viapanam.org.
Villa, Dana R. 'Postmodernism and the Public Sphere.' *American Political Science Review* 86, no. 3 (September 1992): 712–21. https://doi.org/10.2307/1964133.
'What Is the Arab Spring, and How Did It Start?' *Al Jazeera*, accessed 3 October 2023. https://www.aljazeera.com/news/2020/12/17/what-is-the-arab-spring-and-how-did-it-start.
Young, Iris Marion. *Inclusion and Democracy*. Oxford: Oxford University Press, 2000.

12 Technology and the future of democracy

Postcolonial and more-than-human design

In her 2023 book *Decolonising Design: A Cultural Justice Guidebook*, Elizabeth 'Dori' Tunstall proposes some guidelines to 'decolonise' Western research institutions, based on her experience as Dean of Design at the Ontario College of Art and as the first Black woman to occupy this position anywhere in the world.[1] Western designers, as she explains, must overcome the bias that there is a hierarchy of design, where Europe is at the top, China and India in the middle, and everyone else at the bottom.

In 2008, the 'rights of nature' were included in the Ecuador constitution, under the presidency of Rafael Correa, at the instigation of the Pachamama Alliance, a San Francisco-based citizen organisation dedicated to preserving the Earth's tropical rainforests and empowering local indigenous people and in the light of the ecological vision of the Quechua population of the Andes.

These two examples show two more important directions for design for democracy: first, reinforcing the reflection on the colonial aspects of existing digital technologies and supporting decolonial projects and design practices (post-colonial democracy); second, extending the understanding of democratic representation to cover the interests and possibly the rights of the natural environment ('more-than-human' democracy), and moving from 'human-centred' to 'more-than-human' design.

12.1 Post-colonial democracy

The critique of a deliberative democracy model relying too much on argumentative deliberation in the formation of the public sphere is particularly relevant from a global perspective. In fact, it has been argued that Western philosophy-style argumentative deliberation – despite all its merits – cannot and should not be considered as the unique, universal form of democratic communication, to which all people in all countries should adhere.[2] The modes of democratic participation may be different in different regions of the world. It is certainly possible and legitimate that

DOI: 10.4324/9781003303244-16

people in non-Western countries sometimes want to embrace the language of argumentative deliberation typical of the Western (philosophical) tradition, possibly to break the oppression of local social groups who appeal to tradition and people's affective commitments to maintain their oppressive power.[3] At the same time, it is also unreasonable to expect that everybody will be able and willing to do so. The critique of argumentative democracy here connects with the issue of *decolonisation*.

Decolonisation can be seen as an intrinsically democratic ideal on a global scale. In its narrower sense, it concerns the political independence of countries from their (Western) masters. In its broader sense, it means a fight against all cultural factors 'which delimit the participation of colonized people and life forms in the global humanity of life.'[4] Decolonisation requires to undo, as it were, both colonisation and its legacy; or, more precisely, colonisation and the thinking that made it possible and widely accepted: the thinking of (Western) culture as being superior. This broader, cultural, and epistemological approach, inspired among others by the work of Franz Fanon,[5] is sometimes referred to as 'decoloniality.'[6]

In relation to technology and AI, scholars interested in decoloniality note that even well-intentioned initiatives in the so-called AI ethics tend to take for granted Western ethical frameworks as a starting point to promote ideals such as 'AI for Good' or 'Fair and Responsible AI.'[7] Instead, a globally democratic and inclusive ethics of technology must be 'decolonial.' Rather than assuming a universal ethics modelled on the Western tradition, it should foster a more 'intercultural' and 'pluriversal' ethics, where local traditions and approaches to technology coming from the 'AI peripheries' can also influence those coming from the 'AI metropoles.'[8]

12.2 Decolonising AI

12.2.1 Exposing neo-colonial technologies

The current production of AI crucially depends on the extraction of materials, labour, and data from vulnerable, colonised countries and people. This structural dependence is often not sufficiently known and visible. In her book *Atlas of AI*, Kate Crawford makes a systematic review of the many ways in which the production of AI relies on extractive practices from developing countries.[9] In addition to the well-known ingestion and grinding of big amounts of data, AI crucially depends on the extraction of minerals to power the electronics and the consumption of massive amounts of energy to run the machines and water to cool down the data centres. It is difficult to get a precise measure of how much resources are being consumed, but some studies have tried to estimate the environmental impact of the model training process for natural-language processing, the technique which allowed the remarkable advancements of the GPT models (of which ChatGPT is the latest instalment). Over its lifetime, the costliest model, called BERT, consumed the equivalent of a trans-American round-trip by airplane for one person. Building a scientifically respectable model requires training some 4,789 models over a six-month period.[10] Water consumption is probably a less known aspect of AI, but recent studies show that 'training GPT-3 in Microsoft's state-of-the-art U.S. data centers can directly consume 700,000 liters of clean freshwater (enough for producing 370 BMW cars or 320 Tesla electric vehicles).'[11]

The colonial structure is reproduced: while AI is mostly produced by companies in industrialised countries and for those countries' consumers, the mineral extraction, energy production, and water consumption typically affect developing countries more dramatically. Same with labour. The microlabour necessary to train, for instance, ChatGPT is mostly outsourced to companies in Africa who hire workers for miserable salaries and in poor working conditions.[12]

There are also cultural costs for indigenous communities. As explained in the next section,

> Only the most common languages have enough speakers – and enough profit potential – for Big Tech to collect the data needed to support them. Relying on such services in daily work and life thus coerces some communities to speak dominant languages instead of their own.[13]

A first step to resist this neo-colonial trend of AI is documenting it, making it visible, and sparking discussions around it. Crawford did so in her book *Atlas of AI*.[14] In a previous project in collaboration with Joler, one that resonated well with the idea of Critical Design, Crawford had created a work called 'Anatomy of an AI system: The Amazon Echo as an anatomical map of human labor, data and planetary resources.'[15] It is a visual map of the massive network of material and human resources utilised during the life cycle of a single Amazon Echo smart speaker. The work consists of a website, digital publication, physical publication, and digital map. Interestingly, the digital map has also been donated to museums like the MoMA in New York and the London Design Museum, which have created large-scale installations displaying Crawford's and Joler's map.

12.2.2 Using AI for indigenous emancipation

Another way to push back the neo-colonial character of AI is devising projects that support local communities, do not exploit their resources, and support their so-called 'data sovereignty,' which is their ability to remain in control of the data they produce, thereby preventing the appropriation and exploitation by big companies. One example is the project to revitalise *te reo*, the Māori language, by Peter-Lucas Jones and Keoni Mahelona in the village of Kaitaia, New Zealand.[16] Since the British domination of *Aotearoa* – then renamed New Zealand – started in 1840, the local *te reo* language was gradually and forcefully replaced by English. Following more than a decade of initiatives and protests by Māori community leaders, only in 1987 *te reo* was declared an official language. Also, radio stations got funds to publicly broadcast in *te reo*. The manager of one of these radios, Te Hiku Media, was Peter-Lucas Jones. After 20 years of broadcasting, Jones and his life partner, Keoni Mahelona, decided to create a digital database with the audio recording accumulated by the radio. They thought this would help pass on the language to next generations. After asking the local tribes for permission to use their voices, they decided not to upload the database on Facebook or YouTube, to avoid giving Big Tech the license to reuse and exploit the data. So, by relying on Mahelona's engineering background, they build their own platform.[17]

Recently, in 2016, Jones devised a new project: interviewing elderly native *te reo* speakers to preserve their voices and memories. To create a tool for displaying a transcription of the interviews, given the shortage of younger *te reo* speakers whose language skills were good enough to do the manual transcription, he considered natural-language processing, that is, as it were, teaching computer to speak Māori. That looked like a more approachable challenge, as he only needed to find people whose language skills were good enough to train the program by reading a series of standard sentences containing the full range of sounds in Māori, rather than listening to and transcribing the existing audio. They organised a competition and collected 310 hours of recordings in 10 days. By using an existing open-source model from the Mozilla Foundation, they eventually created the first *te reo* speech recognition model. More projects would follow, together with the establishment of new data sovereignty protocols, to prevent people outside the community to use the data for their purposes.[18]

12.2.3 Decolonising design

Decolonisation is also a broader cultural process and using AI to support individual emancipation projects for indigenous populations may not be enough. AI itself should be decolonised, that is, the colonial values and processes behind its production should be exposed and challenged in a more systematic way.

First, there are persisting (colonial) power relations in the design process. These are still, by and large, led by people from richer countries, companies, and universities. Even when genuinely committed to open the design process to the participation of people from different countries and cultures, in the form of stakeholder participation in a Value-Sensitive Design Process or other forms of Participatory Design, wealthier researchers may not be able to transfer sufficient power to others, with the result of people from minority groups contributing to the development of AI projects without having a real influence in directing and benefitting from it.

The Design Justice movement tries to define new methods to ensure that underrepresented and marginalised groups get a real opportunity to contribute to the design process.[19] One important starting methodological point is aiming to develop existing practices and expertise of these groups rather than trying to involve them in processes that largely reflect the design interests and expertise of mainstream groups. At a deeper level, the challenge is that of contesting the alleged universalism of the experiences and the epistemologies of the Western designers and creating concrete methods and tools for letting alternative forms of knowledge and experiences to emerge and impact the design process.

Literature in Participatory Design – in the context of the workplaces – has already emphasised the importance of creating what they call a 'third space,' a hybrid, fertile environment where workers and software professionals can combine their knowledge and getting new insights and plans for action.[20] Participatory design in a post-colonial context requires some additional steps. A recent multidisciplinary project in community-based participatory design and anthropology has studied decolonial design practices in post-colonial Namibia.[21] The project focused on the role of younger generations, the so-called 'born frees' as they were born after the end of colonisation by South Africa. Researchers noticed the importance of creating a new kind of space – the 'safe space' – for thoughts, actions, and mutual learning that would work in political and contested contexts, where people can negotiate and learn from each other about their colonial past. The experience and understanding of colonisation and decolonisation processes by younger Namibians turned out to be different and often in tension with the ones of the older people who fought for liberation and were now involved in governing the country. Speculating and making plans about their political and technological futures, so the researchers argue, require them to first create a space to reflect and engage with each other about their histories and pasts. In the next phase of the project, local and international researchers and design students would, so to speak, enter the safe space and start co-designing tools and prototypes to shape imaginaries and negotiations about possible futures. In the final phase, the safe space would be open to the public through an exhibition in a local museum and an exhibition in virtual reality open to a global audience. This example shows that decolonising AI requires a closer collaboration between designers and anthropologists, to ensure that design practices are supported by an expert engagement with local, contested knowledge and histories.

Finally, beyond individual projects, designers must cultivate tools and methods for systematically decolonising their own design thinking and practices. They must somehow re-educate themselves. In the book *Decolonising Design,*[22] Dori Tunstall, the Dean of Design at Ontario College of Art, and the first Black woman to occupy the position of Dean of a Design faculty anywhere in the world, tells about her experiences in working to decolonise a Western research institution. She recounts the difficulties for Western designers to 'unlearn' some deep-seated biases guiding their profession and embedded in their institutions, research projects, and educational curricula. For instance, the bias that technological development is improving people's lives everywhere, whereas, as seen, it worsened the lives of many people, especially indigenous populations. Also, there is a biased hierarchy of design, where Europe is at the top, China and India in the middle, and everyone else at the bottom. Fighting these biases requires learning more about the work of Black, indigenous and People of Colour communities and reclassifying what is often seen as 'craft' into an actual part of the history of design. It calls for reprioritising the allocation of existing research resources to promote projects that support a redistribution of power and visibility. It also means being able to recognise the forms of marginalisation and colonisation specific to different contexts. Where in Europe an urgent issue is the empowering of people from its former colonies, in India it concerns supporting the Dalit and Adivasi, and in China minority ethnic groups: 'Everywhere has been affected by colonisation, internal or external, so where there have been indigenous peoples displaced from their lands, there is decolonial work to be done.'[23]

12.3 More-than-human democracy

Different philosophical and social science traditions have in the past few centuries criticised and problematised one central element of liberal-democratic philosophies: their reliance on one specific understanding of human agency, coming from Western modern moral philosophy. The human agency of Western moral philosophy is that of an individual human being, taken in abstraction from their material, social, and cultural context, whose actions are described and assessed mainly in relation to the agent's individual mental states such as consciousness and intentionality. That individual is the subject of moral and legal duties and responsibilities, the bearer of moral and legal rights, the agent of democratic deliberation, the ideal signatory of the social contract in the 'State of nature,' the bearer of economic interests in the market, etc.

Philosophers from many different traditions have strongly criticised such an account of human agency, among other things, for underestimating the importance of material, social, and cultural factors in the construction of the moral and political subject; for ignoring the so-called relational nature of human agency, that is its being irreducibly grounded in the connection with other agents. One criticism of the modern liberal philosophy that is very relevant in the present-day philosophy of technology and design is that against a too simplistic distinction between human agents as 'subjects' – of knowledge, of action, and of legitimate interests – and all the rest of the universe as passive 'objects.'[24] The next two sections elaborate on this cricitism.

12.3.1 Object-oriented democracy

In Chapter 2 and across the book I have quoted Langdon Winner's famous slogan: 'artifacts have politics.' This could also be translated into: 'artefacts do things' or 'objects have agency.' In Winner's iconic example, an overpass can enforce a racist policy by preventing public transport and their prevalently Black users to reach Long Island beach. One would be tempted to say that this is just metaphorical talk and that in reality the overpass is only a tool in the hands of the policymakers: it is they who ultimately prevented Black people from reaching Long Island. In fact, the concept of the overpass as a tool sounds more like a metaphor: can any individual person materially *use an overpass as a tool*? It's more like complex institutional choices are embedded in and acting through the material world, in this case the built environment. Philosopher of technology Peter-Paul Verbeek – one of his books iconically titled *What Things Do*[25] – has tried to explain the paradox away. Both human agents and material things have agency; to be more precise, none of them has agency taken in isolation, but both have, through their relationship. The distinction between subjects and objects is less neat than modern philosophers allow for, as human agents and material things constantly influence and shape each other, and agency is the result of their dynamic interaction, where human beings and material things are always at the same time subjects and objects of each other's agency. As summarised in the title of a famous book of Sociologist of Science and Technology Bruno Latour: *We Have Never Been Modern*, by which he means that a sharp distinction between subjects and objects has never existed outside philosophy books. Or, as Winston Churchill once reportedly said, in a plea for the reconstruction of the House of Lords bombed out in 1941: 'We shape our buildings; thereafter they shape us.'

Nowadays there is much talk about the agency of AI systems, but this is not completely new, and AI just made one existing phenomenon more dramatically visible: the agency of so-called socio-technical systems. AI-driven computer systems act and interact in complex networks with human agents and their institutions but so do many other systems: cities, political systems, corporations, and workplaces.

The focus on the role of artificial agency – of artefacts and institutions – in the shaping of human agency is one source of what has been called the 'decentring of human agency.'[26] Accepting the decentring of human agency and the problematisation of the liberal subject of Western modern moral philosophy must *not* entail the elimination of human agency and the refusal of the moral importance of the Western modern-philosophy-based liberal-democratic subject. Human agency still exists and human individual subjects still deserve moral and political respect, incommensurably more so than artefacts. In their relationship with artificial agents, humans remain the only subjects of *moral* agency, responsibility, duties, and rights.

That is why, rather than talking of 'post-humanism,' it is more appropriate to talk of 'more-than-humanist' ethics and politics.[27] The humanist perspective remains important but must be integrated and enriched. In particular, to understand human ethics and politics, including democracy, we must always consider the human subjects and their agency as intrinsically mediated by the agency of artificial systems – artefacts and institutions. It is in this interaction that human agency and values are realised and protected or rather denied and endangered. Another evocative concept by sociologist Bruno Latour is 'Object-Oriented-Democracy.'[28] A focus on objects or artificial systems is crucial to understand the way they (do not) embed, (do not) carry, and (do not) support (human) norms, values, and principles. This awareness has important implications for the design perspective, as already seen in the discussion of Value-Sensitive Design and Critical Design, and as I will also further discuss below. Objects and systems can be designed to support democratic values (through Value-Sensitive Design) and to spark and support democratic debate and engagement about (un) desirable present and future states of affairs (through Critical Design).

12.3.2 The environment's call

A more recent, major contribution to the development of the idea of more-than-human ethics and politics, and, more specifically, what we may call 'more-than-human democracy' has been given by two important, partially related, facts: the environmental crisis and the already-mentioned rise of decolonial thinking.

On the one hand, it has become evident that an exclusive focus on the rights and interests of human beings only conceived as individual agents and bearers of individual interests is not only insufficient but also nefarious for the preservation of the planet. It may be argued that this exclusive focus has in fact played an important role in creating the environmental crisis of our time. The environment has never been sufficiently considered as a legitimate actor in the mainstream Western ethical political debate. In this sense, even the most advanced and progressive ethical and political theories and practices for the inclusion of all human beings in an ideal global democracy would be insufficient. The planet itself – as a bearer of legitimated interests – must somehow be included in a future global democracy.

On the other hand, a more-than-human approach has been promoted by decolonial thinking. First, decolonial thinking, as seen above, fiercely opposes not only the political domination intrinsic to coloniality but also its legacy or, better, the culture that produced and supported that political domination ('coloniality'). As a cultural phenomenon, colonisation existed before and continues to exist after the end of colonisation as a political phenomenon. Decolonial thinkers see colonial culture as a combination of two elements: (a) a strong capitalistic imperative to indefinitely push economic growth through the continuous, unlimited extraction and conversion of natural resources around the planet, and (b) a racist belief in the right to systematically exploit the natural resources and the people in regions inhabited by people considered as inferior, in the alleged legitimate interest of promoting the well-being of the people of the colonising countries considered as superior.

The racist belief has arguably supported the capitalist colonial exploitation of natural resources in a more subtle way. It has prevented colonial cultures to recognise and value local, indigenous cultures and institutions, often considered primitive, inferior, or even non-existing. Local cultures are sometimes more respectful of the environment, and attending to their values may help prevent colonial countries from harming the environment. It might have also offered them some precious insights for the establishment of a more balanced co-existence between human and non-human agents.

Consider the practices of mourning plants and animals in the indigenous Marind communities of rural Merauke West Papua, studied by the French and Chinese environmental anthropologist Sophie Chao.[29] Marind communities have built practices to memorialise the radical loss of non-human lives and relations caused by capitalist-driven landscape transformation, such as deforestation. These practices include the ritual weaving of sago bags as a form of collective healing, and the creation of songs prompted by encounter with animal roadkill. The so-called 'multispecies mourning' is also a form of political resistance, a creative refusal of the capitalist logic of extraction and a reaffirmation of the importance of 'multispecies solidarity.'[30]

These approaches have recently been increasingly advocated for also in more explicitly legal and political terms. The most evident success has probably been the inclusion of the 'rights of nature' in the Ecuador constitution, achieved in 2008 under the presidency of Rafael Correa, at the instigation of the Pachamama Alliance, a San Francisco-based citizen organisation dedicated to preserving the Earth's tropical rainforests and empowering local indigenous people,[31] and in the light of the ecological vision of the Quechua population of the Andes, which is centred on community and harmony with nature.[32]

To be sure, we should not indulge either in a romantic idealisation of non-Western cultures, as always being respectful of the environment and constantly living in harmony with it. This would be a distorted representation of reality. The point is to criticise the colonial attitude that prevents (Western) cultures to recognise and appreciate the existence of other cultures, respecting their values and principles, and possibly even reflecting on how their insights may help also people in the West build a more balanced relationship with the environment.

12.4 More-than-human design

12.4.1 Beyond human-centred design

An increasingly popular formula among digital technology designers is that of 'human-centred design.' From a technical perspective, this generally refers to including more psychological studies in the design process to ensure that products are aligned with the actual skills and capabilities of the users, or including the potential users in the design process to ensure this alignment.[33] The human here is the potential user. Imagine a designer trying to design a partially automated car that responds to the will and skills of potential drivers or users. From an ethical, legal, and political point of view, 'human-centred design' often refers to the idea that technological systems should be centred around human needs, values, and rights as opposed to be designed for the sake of mere technological advancement, a generic strive towards the value of innovation, or, more commonly, the profit of the companies producing it. 'Human' stands here for human values in contrast with 'technological pushes' or 'corporate profit.' This sense is closely related to Responsible Innovation and Value-Sensitive Design.

Taking a *more-than-human* approach may help see the limits of 'human-centred' design.[34] One first limitation of this type of design approach is that it sees the human element as something separate from technology and design, something to which design should adapt. In the schematic modern philosophical approach criticised by Latour: producing objects for subjects. Sociology of science[35] and post-phenomenological approaches in philosophy of technology,[36] among others, have convincingly argued that this picture is inaccurate and misleading. Human values and technology tend to shape each other dynamically and continuously. Rather than taking human values, needs, etc. for granted and designing products around them, the design process should create spaces, objects, and systems that make people and communities gather, debating and developing awareness on their relationship with technology and producing new visions and values for possible future interactions with technology. Following on the example of the car design, imagine a designer who creates scenarios, objects, and activities to let groups of people debate over their relationship with cars, road traffic, transportation, time, and the environment. This is the main idea of Critical and Speculative Design, aiming at engaging communities in deliberation and creating publics or civics to imagine and plan for possible futures as opposed to just involving stakeholders to design new products.[37]

12.4.2 Beyond sustainability

A second, important limitation of the human-centred approach is its focus on humans in separation from their natural environment. One can certainly integrate the human-centred design with a focus on sustainability as one of the values to be designed for. However, this would be insufficient. In one of its original formulations of the UN Brutland Commission, sustainability requires to 'meet the needs of the present without compromising the ability of future generations to meet their own needs.'[38] Human needs are still at the centre, and the environment is mainly considered as a source of benefits for present and future generations. Sustainability is still a human value. A more-than-human approach would rather consist in seeing the humans and the environments as actors in interaction, as an ecosystem where non-human elements also have legitimate needs to survive and flourish. Design scholar Laura Forlano claims that designers 'can operate as mediators of these [non-human] forces and advocates for less visible nonhuman stake-holders.'[39] By creating spaces, objects, and experiences that could radically 'shift our experience of the world' as just a bunch of functional objects and resources to be used by us, and 'dramatically reevaluat[ing] our 'needs' and our requests to government and corporations.'[40]

Consider as an example smart cities, which are urban spaces where data-driven applications help manage and optimise new services and activities such as transportation services that adapt to the traffic conditions, energy platforms that adapt to the users' consumption patterns, and municipality services that are crucially mediated and made more efficient by the acquisition and elaboration of data about the citizens. While designing services for smart cities, designers may take a *market-driven, individualistic approach* and see people as mainly users or consumers who want to fulfil their daily needs and wants: designing for faster transportation, cheaper energy, more efficient administration, etc. In addition, they may take a *socio-technical and political approach* and see people as citizens who want to engage in a reflection on how different socio-technological projects could impact their and their fellow citizens lives: how would the city look like, how would burdens and benefits be distributed, how would people's lives and relationships with the urban space change, what would be the consequences of the vast scale of data acquisition on different groups of citizens' freedom and power vis-à-vis governments and corporations, etc., designing for political engagement around the smart city. Finally, designers may see *people, technology, and the city as part of a broad ecosystem*, in which multiple human perspectives must co-exist and different human and non-human agents must live and flourish, for instance, in one example based in Australia: by including the migratory patterns of wildlife and indigenous knowledge systems and cultures of managing land in the design process.[41]

Design practices cannot solve the environmental issues alone; this requires a deeper political and economic change. But they can and should help people – citizens, policymakers, designers, philosophers, and many others – to develop a better awareness of the interconnectedness between them and between them and the planet.

Notes

1 Elizabeth Tunstall, *Decolonizing Design: A Cultural Justice Guidebook* (Cambridge, MA: The MIT Press, 2023).
2 Iris Marion Young, *Inclusion and Democracy* (Oxford: Oxford University Press, 2000).
3 Elizabeth M. Zechenter, 'In the Name of Culture: Cultural Relativism and the Abuse of the Individual,' *Journal of Anthropological Research* 53, no. 3 (October 1997): 319–47, https://doi.org/10.1086/jar.53.3.3630957.
4 Rachel Adams, 'Can Artificial Intelligence Be Decolonized?,' *Interdisciplinary Science Reviews* 46, no. 1–2 (2021): 182, https://doi.org/10.1080/03080188.2020.1840225.
5 Frantz Fanon, *Black Skin, White Masks*, 1st Evergreen ed. (New York: Grove Press, 1982).
6 Adams, 'Can Artificial Intelligence Be Decolonized?,' 179–81.
7 Adams, 182.
8 Shakir Mohamed, Marie Therese Png, and William Isaac, 'Decolonial AI: Decolonial Theory as Sociotechnical Foresight in Artificial Intelligence,' *Philosophy and Technology* 33, no. 4 (1 December 2020): 659–84, https://doi.org/10.1007/S13347-020-00405-8/METRICS.
9 Kate Crawford, *Atlas of AI: Power, Politics, and the Planetary Costs of Artificial Intelligence* (New Haven: Yale University Press, 2021).
10 Karen Hao, 'Training a Single AI Model Can Emit as Much Carbon as Five Cars in Their Lifetimes,' *MIT Technology Review*, 2019, https://www.technologyreview.com/2019/06/06/239031/training-a-single-ai-model-can-emit-as-much-carbon-as-five-cars-in-their-lifetimes/.
11 Pengfei Li et al., 'Making AI Less "Thirsty": Uncovering and Addressing the Secret Water Footprint of AI Models,' *arXiv*, accessed 18 September 2023, http://arxiv.org/abs/2304.03271.
12 Billy Perrigo, 'OpenAI Used Kenyan Workers on Less Than $2 Per Hour: Exclusive,' *Time*, 18 January 2023, https://time.com/6247678/openai-chatgpt-kenya-workers/.
13 Karen Hao, 'A New Vision of Artificial Intelligence for the People,' *MIT Technology Review*, 22 April 2022, https://www.technologyreview.com/2022/04/22/1050394/artificial-intelligence-for-the-people/.
14 Crawford, *Atlas of AI*.
15 Kate Crawford and Vladan Joler, 'Anatomy of an AI System,' accessed 4 October 2022, https://anatomyof.ai/.

16 Hao, 'A New Vision of Artificial Intelligence for the People.'
17 Hao.
18 Hao.
19 Sasha Costanza-Chock, *Design Justice: Community-Led Practices to Build the Worlds We Need* (Cambridge, MA: The MIT Press, 2020).
20 Michael Muller and Allison Druin, 'Participatory Design: The Third Space in HCI,' in *The Human-Computer Interaction Handbook: Fundamentals, Evolving Technologies and Emerging Applications*, ed. Julie A. Jacko (New York: CRC Press, 2012), 1061–1081.
21 Asnath Paula Kambunga et al., 'Decolonial Design Practices: Creating Safe Spaces for Plural Voices on Contested Pasts, Presents, and Futures,' *Design Studies* 86 (1 May 2023): 101170, https://doi.org/10.1016/J.DESTUD.2023.101170.
22 Tunstall, *Decolonizing Design*.
23 ICoD, 'ICoD Interviews Elizabeth (Dori) Tunstall on Decolonizing Design,' 22 May 2023, https://www.theicod.org/resources/news-archive/icod-interview-dori-tunstall.
24 Bruno Latour, *We Have Never Been Modern*, 3. print. (Cambridge, MA: Harvard University Press, 1994).
25 Peter-Paul Verbeek, *What Things Do: Philosophical Reflections on Technology, Agency, and Design* (University Park, PA: Penn State University Press, 2005).
26 Laura Forlano, 'Decentering the Human in the Design of Collaborative Cities,' *Design Issues* 32, no. 3 (1 July 2016): 42–54, https://doi.org/10.1162/DESI_a_00398.
27 Peter-Paul Verbeek, 'Cultivating Humanity: Towards a Non-Humanist Ethics of Technology,' in *New Waves in Philosophy of Technology*, eds. Jan Kyrre Berg Olsen, Evan Selinger, and Søren Riis (London: Palgrave Macmillan UK, 2009), 241–63, https://doi.org/10.1057/9780230227279_12.
28 Bruno Latour and Peter Weibel, eds., *Making Things Public: Atmospheres of Democracy* (Cambridge, MA: The MIT Press, 2005).
29 Sophie Chao, 'Multispecies Mourning: Grieving as Resistance on the West Papuan Plantation Frontier,' *Cultural Studies* 37, no. 4 (4 July 2023): 553–79, https://doi.org/10.1080/09502386.2022.2052920.
30 Chao.
31 'Ecuador First to Grant Nature Constitutional Rights,' *Capitalism Nature Socialism* 19, no. 4 (2008): 131–33, https://doi.org/10.1080/10455750802575828.
32 Oliver Balch, 'Buen Vivir: The Social Philosophy Inspiring Movements in South America,' *The Guardian*, 4 February 2013, sec. Guardian Sustainable Business, https://www.theguardian.com/sustainable-business/blog/buen-vivir-philosophy-south-america-eduardo-gudynas.
33 Guy A. Boy, *The Handbook of Human-Machine Interaction: A Human-Centered Design Approach* (Boca Raton, FL: CRC Press, 2017).
34 Elisa Giaccardi and Johan Redström, 'Technology and More-Than-Human Design,' *Design Issues* 36, no. 4 (September 2020): 33–44, https://doi.org/10.1162/DESI_A_00612.
35 Latour, *We Have Never Been Modern*.
36 Verbeek, *What Things Do: Philosophical Reflections on Technology, Agency, and Design*.
37 Anthony Dunne and Fiona Raby, *Speculative Everything: Design, Fiction, and Social Dreaming* (Cambridge, MA and London: MIT Press, 2013); Carl

DiSalvo, *Design as Democratic Inquiry: Putting Experimental Civics into Practice* (Cambridge, MA: The MIT Press, 2022).

38 United Nations, 'Sustainability,' accessed 8 October 2023, https://www.un.org/en/academic-impact/sustainability.

39 Forlano, 'Decentering the Human in the Design of Collaborative Cities.'

40 Forlano.

41 Marcus Foth et al., 'A Capitalocentric Review of Technology for Sustainable Development: The Case for More-than-Human Design,' in *Global Information Society Watch 2020: Technology, the Environment and a Sustainable World: Responses from the Global South* (Melville: Association for Progressive Communications (APC), 2020), https://doi.org/10.1080/03066150.2016.1235036.

References

Adams, Rachel. 'Can Artificial Intelligence Be Decolonized?' *Interdisciplinary Science Reviews* 46, no. 1–2 (2021): 176–97. https://doi.org/10.1080/03080188.2020.1840225.

Balch, Oliver. 'Buen Vivir: The Social Philosophy Inspiring Movements in South America.' *The Guardian*, 4 February 2013, sec. Guardian Sustainable Business. https://www.theguardian.com/sustainable-business/blog/buen-vivir-philosophy-south-america-eduardo-gudynas.

Boy, Guy A. *The Handbook of Human-Machine Interaction: A Human-Centered Design Approach*. Boca Raton, FL: CRC Press, 2017.

Chao, Sophie. 'Multispecies Mourning: Grieving as Resistance on the West Papuan Plantation Frontier.' *Cultural Studies* 37, no. 4 (4 July 2023): 553–79. https://doi.org/10.1080/09502386.2022.2052920.

Costanza-Chock, Sasha. *Design Justice: Community-Led Practices to Build the Worlds We Need. Information Policy*. Cambridge, MA: The MIT Press, 2020.

Crawford, Kate. *Atlas of AI: Power, Politics, and the Planetary Costs of Artificial Intelligence*. New Haven: Yale University Press, 2021.

Crawford, Kate, and Vladan Joler. 'Anatomy of an AI System.' Accessed 4 October 2022. https://anatomyof.ai/.

DiSalvo, Carl. 'Design as Democratic Inquiry: Putting Experimental Civics into Practice. Cambridge, MA: The MIT Press, 2022.

Dunne, Anthony, and Fiona Raby. *Speculative Everything: Design, Fiction, and Social Dreaming*. Cambridge, MA and London: MIT Press, 2013.

'Ecuador First to Grant Nature Constitutional Rights.' *Capitalism Nature Socialism* 19, no. 4 (2008): 131–33. https://doi.org/10.1080/10455750802575828.

Fanon, Frantz. *Black Skin, White Masks*, 1st Evergreen ed. New York: Grove Press, 1982.

Forlano, Laura. 'Decentering the Human in the Design of Collaborative Cities.' *Design Issues* 32, no. 3 (1 July 2016): 42–54. https://doi.org/10.1162/DESI_a_00398.

Foth, Marcus, Monique Mann, Laura Bedford, Walter Fieuw, and Reece Walters. 'A Capitalocentric Review of Technology for Sustainable Development: The Case for More-than-Human Design.' In *Global Information Society Watch 2020: Technology, the Environment and a Sustainable World: Responses from the Global South*. Melville: Association for Progressive Communications (APC), 2020. https://doi.org/10.1080/03066150.2016.1235036.

Giaccardi, Elisa, and Johan Redström. 'Technology and More-Than-Human Design.' *Design Issues* 36, no. 4 (September 2020): 33–44. https://doi.org/10.1162/DESI_A_00612.

Hao, Karen. 'A New Vision of Artificial Intelligence for the People.' *MIT Technology Review*, 22 April 2022. https://www.technologyreview.com/2022/04/22/1050394/artificial-intelligence-for-the-people/.

———. 'Training a Single AI Model Can Emit as Much Carbon as Five Cars in Their Lifetimes.' *MIT Technology Review*, 2019. https://www.technologyreview.com/2019/06/06/239031/training-a-single-ai-model-can-emit-as-much-carbon-as-five-cars-in-their-lifetimes/.

ICoD. 'ICoD Interviews Elizabeth (Dori) Tunstall on Decolonizing Design.' 22 May 2023. https://www.theicod.org/resources/news-archive/icod-interview-dori-tunstall.

Kambunga, Asnath Paula, Rachel Charlotte Smith, Heike Winschiers-Theophilus, and Ton Otto. 'Decolonial Design Practices: Creating Safe Spaces for Plural Voices on Contested Pasts, Presents, and Futures.' *Design Studies* 86 (1 May 2023): 101170. https://doi.org/10.1016/J.DESTUD.2023.101170.

Latour, Bruno. *We Have Never Been Modern*, 3 Print. Cambridge, MA: Harvard University Press, 1994.

Latour, Bruno, and Peter Weibel, eds. *Making Things Public: Atmospheres of Democracy*. Cambridge, MA: The MIT Press 2005.

Li, Pengfei, Jianyi Yang, Mohammad A. Islam, and Shaolei Ren. 'Making AI Less "Thirsty": Uncovering and Addressing the Secret Water Footprint of AI Models.' *ArXiv*, accessed 18 September 2023. http://arxiv.org/abs/2304.03271.

Mohamed, Shakir, Marie Therese Png, and William Isaac. 'Decolonial AI: Decolonial Theory as Sociotechnical Foresight in Artificial Intelligence.' *Philosophy and Technology* 33, no. 4 (1 December 2020): 659–84. https://doi.org/10.1007/S13347-020-00405-8/METRICS.

Muller, Michael, and Allison Druin. 'Participatory Design: The Third Space in HCI.' In *The Human-Computer Interaction Handbook: Fundamentals, Evolving Technologies and Emerging Applications*, edited by Julie A. Jacko, 1061–81. New York: CRC Press, 2012.

Perrigo, Billy. 'OpenAI Used Kenyan Workers on Less Than $2 Per Hour: Exclusive.' *Time*, 18 January 2023. https://time.com/6247678/openai-chatgpt-kenya-workers/.

Tunstall, Elizabeth. *Decolonizing Design: A Cultural Justice Guidebook*. Cambridge, MA: The MIT Press, 2023.

United Nations. 'Sustainability.' Accessed 8 October 2023. https://www.un.org/en/academic-impact/sustainability.

Verbeek, Peter-Paul. 'Cultivating Humanity: Towards a Non-Humanist Ethics of Technology.' In *New Waves in Philosophy of Technology*, edited by Jan Kyrre Berg Olsen, Evan Selinger, and Søren Riis, 241–63. London: Palgrave Macmillan, 2009. https://doi.org/10.1057/9780230227279_12.

———. *What Things Do: Philosophical Reflections on Technology, Agency, and Design*. University Park, PL: Penn State University Press, 2005.

Young, Iris Marion. *Inclusion and Democracy*. Oxford: Oxford University Press, 2000.

Zechenter, Elizabeth M. 'In the Name of Culture: Cultural Relativism and the Abuse of the Individual.' *Journal of Anthropological Research* 53, no. 3 (October 1997): 319–47. https://doi.org/10.1086/jar.53.3.3630957.

Index

Made in the USA
Las Vegas, NV
21 September 2026

2c77d589-5fbd-48b2-9220-2b2b18140c33R01